MANUAL

OF

AGRICULTURE,

FOR

THE SCHOOL, THE FARM,

AND

THE FIRESIDE.

By GEORGE B. EMERSON,
Author of a "Report on the Trees and Shrubs of Massachusetts,"

AND

CHARLES L. FLINT,
SECRETARY OF THE STATE BOARD OF AGRICULTURE,
Author of a Treatise on "Milch Cows and Dairy Farming," and "Grasses and Forage Plants," etc., etc.

BOSTON:
SWAN, BREWER & TILESTON,
131 Washington Street.
1862.

[Published under the sanction of the State Board of Agriculture.]

RECOMMENDATIONS.

The Massachusetts State Board of Agriculture, after a careful revision of the work, passed the following Resolution:—

"*Resolved*, That this Board approve of the Manual of Agriculture, submitted by its authors, Messrs. George B. Emerson and Charles L. Flint, and recommend its publication by these gentlemen, as a work well adapted for use in the schools of Massachusetts."

LETTER FROM HON. MARSHALL P. WILDER.

DORCHESTER, November 22, 1861.

Gentlemen,—The first demand of life is for food, and the only supply is from the products of the soil. Agriculture is, therefore, of primary importance, not only as the source from which we derive our daily bread, but as the parent of all the other great industrial pursuits. Without agriculture there can be no commerce or manufactures, no population or prosperity. Every one, of whatever vocation, is interested in its welfare, and every man, woman and child, should have some knowledge of the fundamental principles of this most useful art.

To accomplish so desirable an object, and especially to implant in the minds of youth an abiding love for this honorable employment, the Massachusetts Board of Agriculture have caused the Manual of Agriculture for Schools to be prepared under its own supervision and direction. The task of preparing this volume was confided to Messrs. George B. Emerson and Charles L. Flint. Worthily have the authors performed their duty, and most cordially do I recommend the book as being admirably adapted to the use of schools, and equally valuable to the cultivators of the soil. I take great pleasure in commending it, not only to the people of Massachusetts, but to the farming community throughout our country.

MARSHALL P. WILDER.

Messrs. SWAN, BREWER & TILESTON.

OPINIONS OF THE PRESS.

[From the Montreal Transcript.]

One of the most useful books of the kind we have ever met with.

[From the Ohio Farmer.]

The plan of the work is excellent and the matter equal to the plan. Had we the power we would have it taught in every school in America.

[From the Barnstable Patriot.]

It may be safely accepted as a standard work upon all that it professes to teach. We recommend it to all our farmers as a most useful hand-book.

[From the Gazette, Montreal.]

The plan of the work is admirable, and the writing has the merit of being both clear and concise.

PREFACE.

This book is intended to supply an important defect in the instruction of youth. Children should not only receive instruction in the various studies now pursued in the schools, but they should be enabled to obtain the rudiments of a correct agricultural education in the forming period of life. Many of them are destined to work upon the surface of the earth, and to use the materials of which it is formed. They are to coöperate with the great powers of nature and to be able, in many instances, to control these powers. They instinctively long to become acquainted with the materials they are to work upon, and the powers they are to work with. An agricultural education should, therefore, be commenced in childhood.

Children are always interested in animals and other natural objects. There is scarcely a fact exhibited in nature which does not, at once, interest their curiosity and appeal to their imagination. They should, therefore, learn about the earth, the different kinds of soil, the names of the different kinds of rocks and their uses. Tell children that, in the soil, the roots of the grasses dissolve the stones, and carry particles of them up and leave them, infinitely minute but infinitely numerous, in the stem and leaves, and that these particles make the grass strong enough to stand up, and they will be interested in the information.

Children should learn the appearance and properties of every common metal; for there is no person to whom the knowledge would, in any part of life, be useless.

There are seventeen, perhaps nineteen, elementary substances in all, which enter into the composition of plants and animals. These, combined, form the numberless objects which are exhibited by the vegetable and animal kingdom, and children should be taught the nature, properties and uses of these elements.

We live surrounded by the air, which is composed of two invisible gases, oxygen and nitrogen, both essential to the life of every animal and of every plant. Children do not see air, nor oxygen, nor nitrogen; but they are just as able to understand this mixture, both ingredients of which are invisible, as they are to understand, what they often see, that salt becomes invisible in water, and steam and smoke in the air. Show them a piece of clean, bright iron, and another of rusty iron, and explain to them that it is the oxygen of the air which has combined with the iron, and converted it into rust or dirt, and they will be prepared to comprehend all that you have to teach them about the combinations of oxygen and other elements with each other.

Plants feed on carbonic acid and ammonia. When children understand what these are, there is nothing you can tell them more curious and wonderful than the fact, that the wind which blows from the habitations of men carries with it these gases, offensive and poisonous to animals, but that rain dissolves and brings them down to the roots, and that plants live upon them.

All these facts are perfectly intelligible to children at an age as early as that at which they are capable of learning grammar and geography. Every fact to be presented is a simple fact. There is scarcely one in natural history, or in the sciences on which the knowledge of the principles of agriculture is founded, which is not as easily comprehended as any of the ideas of history. Which, for example, is easier for a child to comprehend, oxygen and its action, or civil government, nitrogen or confederation, carbon or a league, phosphorus or the reformation? Which will a child be most likely to understand and be interested in, the little root and seed leaves of a plant just up, and the future plant rising between them, or a convocation of ambassadors to consider the Edict of Nantes, or a plenipotentiary to protest against the Solemn League and Covenant?

So far is a knowledge of the powers which are in operation in nature, the action of heat and light, and the causes of wind and rain, with their effects upon the vegetable kingdom, from being difficult, it is that which every sensible child instinctively longs for and delights in; and as to its being speculative, it is the only knowledge which is absolutely sure to be useful to every person who obtains it.

An essential part of good education is admitted to be the discipline of the faculties. The faculties which come earliest into play, are suited to observe and learn the facts presented in nature. These facts and objects are, therefore, the proper, natural study of the earliest years of children.

We would, therefore, gladly make a knowledge of the principles which underlie an intelligent understanding of the art of agriculture, the basis of education, especially for all those who are destined to the happy fortunes of living in the country. And who is there, that has ever lived in the country, who does not hope, some day or other, to live on his own farm or among the farms of his friends, in the midst of the things which God has made?

The State Board of Agriculture, strongly impressed with the importance of these views, has caused this volume to be prepared as a text-book for schools, with the hope that it may do something to lay the foundation of a complete agricultural education, where it may most effectually be done, in the district school.

In the execution of the work, Mr. Emerson has prepared the first thirteen chapters and the twenty-first chapter, upon the Rotation of Crops, and Mr. Flint the remainder, commencing with the fourteenth chapter. Many of the more important principles embraced in the topics discussed, have been repeated in various forms and in different connections, for the purpose of impressing them more strongly upon the mind, but it is confidently hoped that this fact will not make the volume unattractive to the general reader.

The authors do not lay any claim to originality. They have availed themselves of the information of scientific and practical men, and have tried to state it in a concise and attractive form, so far as the subjects treated of seemed to make it practicable to do so.

Boston, November, 1861.

MANUAL OF AGRICULTURE.

CHAPTER I.

INTRODUCTION.

1. Agriculture is the art of cultivating the earth. It includes whatever is necessary for finding out the nature of the soil, clearing up the land, rendering it healthy, and preparing it for tillage, and ploughing it, and the sowing, weeding and harvesting the crops.

2. The object of agriculture should be to enrich the earth, and make it produce the largest crops, of the greatest value, at the least expense of land, time, and labor.

3. In order to attain this object, the husbandman must have capital,—that is, money, for the necessary expenditures; labor, or hands for the operations required; knowledge of the best ways of working; and intelligence, in order to direct the application of the capital and labor.

4. A complete farm ought to have woodland, pasture land, meadow or grass land, arable land, an orchard, a garden spot, and space for roads.

It should have a farmer's house, a barn or stable for horses, oxen, sheep, and swine, and for crops, a tool-house, a dairy, fences, walls or hedges, and wells or springs.

It would be desirable to have a stream running through it or by it, and to have a pond or swamp connected with or belonging to it.

5. A husbandman also wants capital to stock his farm with cattle and other animals, and to furnish it with carts, wagons, ploughs, and other tools.

6. To carry on a farm successfully, a good deal of knowledge and a high degree of intelligence are necessary, and these are to be obtained partly by study, and partly by practice.

By *study* the farmer should find out—1st, the nature and mode of growth of the plants and animals he is to have to do with; and 2d, the nature and properties of the soil and of the atmosphere on and in which they live.

Practice, or experience, is acquired by doing himself the work on a farm, under the guidance of a skilful farmer. By means of both study and experience, he may learn to avail himself of all the means of improving his farm which are in his reach, or which he can bring within his reach.

7. The farmer, indeed, should have that exact knowledge of facts and principles, of effects and their causes, which is called **Science.** For example, if a farmer knows exactly what a plant is made of, and what nourishment it requires, and whether a particular soil contains the substances which will nourish that plant, and, if it do not, knows exactly what kind of manure does contain proper nourishment for the plant, that farmer has a scientific knowledge of the plant, of the soil, and of the manure. He has the science necessary to the culture of that plant. Science is exact knowledge, obtained by

the observation and experience of many observers, and its natural fruit is "the substitution of rational practice for unsound prejudice."

8. You see then what is the *use* of a scientific knowledge of the principles of agriculture. It prepares a person for the practice of agriculture. A person who has thoroughly learned the scientific principles, will understand, without any difficulty, the reasons for the operations of agriculture.

9. But science will not be sufficient without practice. Practice teaches a thousand things which have not got into the books. But a knowledge of scientific principles opens one's eyes to observe and see many facts which the more unenlightened laborer cannot see, and to perceive the connection between facts which to the ignorant person seem to have no connection.

10. The farmer, therefore, should have a good education. For no one is more highly benefited by a good education. The farmer pursues one of the most important occupations in the world. Almost all the food of civilized man is produced on the farm. The quantity and excellence of the food thus produced depend upon the skill and intelligence with which the farm is managed. Nothing can be done so well by an ignorant and careless person, as by a person of intelligence and knowledge, and there is no place where knowledge is more important than it is on a farm.

11. Of the value of exact knowledge to a farmer there is abundant evidence. Such progress has been made, within a few years past, in the various arts which belong to agriculture, that the produce from the farms in many parts of Europe, particularly of England, is twice as great, on the same land, and with the same amount of

labor, as it was thirty years ago. Now, the improvements which have been made on English farms may be made on American farms, by the use of the same means.

12. Those means are the application of science to the treatment of soils, manures, modes of tillage, and management of animals; and improvements in the various tools and machines used in the work of farming. And no person can wisely make this application, and avail himself fully of these improvements, who is not well educated.

13. Besides, we have evidence nearer home of the value of knowledge to a farmer. The farms in New England, which have been conducted with intelligence, knowledge, forethought and economy, have, in many instances, made, out of poor men, men well to do in the world, and rich enough to command all the comforts and enjoyments of life. Many of the towns in Massachusetts which have been always wholly devoted to agriculture, are among the most thriving towns in the State.

14. But, the question will be asked, suppose a farmer to be well educated; will he have time to keep up his knowledge?

If a farmer have the good fortune to obtain a good education in his early years, he will have more time and stronger inducements to keep up and add to his knowledge, than almost any one else. One peculiar advantage in the occupation of a farmer is that, while it gives full exercise to the powers of the body, it leaves time, at least in this country, for a very full exercise of the powers of the mind. Every operation on the farm calls into use the farmer's knowledge and intelligence; and the long evenings of one half of the year give him ample time for reading and thought. Watching the nature and action of scientific principles will give a new interest and pleasure to every operation in which the farmer engages; and his success in their application will furnish a strong motive for new acquisitions.

15. There is no doubt that men of science are liable to make mistakes, partly because their science is not thorough enough, and partly because very much of what is most important can be learned only by one's own observation. It is the *union* of science and practice which alone can make a perfect farmer.

16. It is often supposed that the scientific principles necessary for intelligent farming are difficult to be understood. But this is very far from being the case. What chemistry teaches about air, water, arable soil, the nature of plants, manure, and what it is made of, is so easy to be understood, that every well-informed teacher may, in a dozen lessons, and with the simplest means of instruction, impart to the commonest farmer's boy an accurate knowledge of it.

17. The learning these things will make the difference between ignorance and knowledge, between seeming stupidity and real brightness. It will be a great benefit to the individual and to the country. The boy who has been taught in school on what the fertility of the soil depends, and the great danger of the land's being worn out in consequence of wasting the most valuable kinds of manure, and who has been told by his teacher that he who wastes the conditions of fertility is guilty of an offence against the poor, against himself, and against society, will certainly, when he grows to man's estate, see how important it is that nothing essential to fertility shall be lost, and will take the greatest pains to save and to use every thing which is thus valuable.

18. What is chemistry? It is the *science* which tells us what water, air, soil, and all other things are, what they are made of, and how the elements of which they are made act upon each other; and a person who studies these things, and makes experiments upon them, is called a *chemist.*

CHAPTER II.

THE AIR AND THE GASES IN IT.

19. **The Air** is that which we breathe, and by which we are constantly surrounded. It is very thin and light, and yet it has some little weight. We cannot see it, and yet it is always about us and touching us. The wind is air in motion. We feel the wind, and we may feel the still air when we move our hand rapidly in it; and we also feel and hear it when we move a stick swiftly through it.

If I fill a bladder with air, and press it, the bladder yields; but as soon as the pressure is withdrawn, it swells out again to its former size. This is because the air is springy or *elastic.* It is essential to burning, or *combustion.* Without air, the candle would be extinguished, and the fire would go out. It is not less necessary to the life of man and other animals, and to plants.

20. The air is composed of a thin fluid or gas, called *oxygen,* (which means, producer of acids,) mixed with another air or gas called *nitrogen,* (producer of nitre,) or *azote,* (not sustaining life.) The air also contains a gas called *carbonic acid,* a small but variable quantity of

watery vapor, and commonly has floating in it smoke and dust, and minute portions of various gases which serve as food to plants, the most important of which are ammonia and sulphuretted hydrogen.

21. **Oxygen** is the vital part of the air—that which is essential to our life, and also to combustion. It is *invisible*, and has no taste or smell. Oxygen is thought to be a *simple substance;* that is, no person has ever succeeded in showing that it is a mixture or compound of any two substances. It is therefore called an **Element,** or elementary substance.

It is one of the most abundant and widely diffused substances known. It forms eight parts out of nine, by weight, in the composition of water. It enters into the composition of nearly all the rocks and different kinds of earth, and is one of the constituents of all portions of the bodies of plants and animals.

22. A considerable portion of every known rock is oxygen, combined with some other element. How it got into the rocks we do not know. Oxygen has a strong tendency to penetrate into every thing; it has a great attraction for iron, copper, lead, and most of the other metals, and for nearly all the other substances of which earths are composed, and combines with them intimately, and completely changes their appearance and properties. Iron left for any time in moist air *rusts*, or is gradually covered with a dirty reddish substance, which we call *rust*, which is made up of oxygen and particles of the iron with which it has united. This the chemists call *oxide of iron.* The iron has been *oxidized.*

This oxide of iron is often found in the earth in great quantities, forming a brownish, heavy dirt or earth; sometimes beautiful rocks or ores. Similar earths or minerals are formed by oxygen uniting with other metals. These compounds are called *oxides*.

23. Oxygen was called a producer of acids, because it is an element of many of the most powerful acids; and the name *acid* is given to several substances which are extremely sour and very corrosive, and produce the effect of turning vegetable blue colors red.

24. Oxygen, for instance, unites with sulphur, or brimstone, in two proportions. If there be sixteen parts by weight of sulphur to sixteen of oxygen, *sulphurous acid* is formed; sixteen of sulphur to twenty-four of oxygen form *sulphuric acid*, commonly called oil of vitriol, a heavy liquid, looking like oil.

25. Eight parts out of nine in the composition of water are oxygen; the other part is hydrogen.

26. **Hydrogen** (water producer) is an invisible air or gas, *elastic*, and without color, taste, or smell, and lighter than any other substance known. One hundred cubic inches of hydrogen weigh $2\frac{13}{100}$ grains.

27. Oxygen, which is a little heavier than common air, is sixteen times heavier than hydrogen.

28. And common air is about 816 times lighter than pure water.

29. **Nitrogen** is a gas which alone does not sustain combustion, nor the breathing or respiration of animals. A burning candle placed in a vessel full of it goes immediately out. An animal placed in it immediately dies. It is not supposed to be poisonous, but merely inert. It serves to temper the violent action of oxygen, which, without it, might consume the lungs which should breathe it. It enters as an essential element into the structure of animals and plants. It has neither color, taste, nor smell.

30. But it is only when alone, or when merely mixed with oxygen, as in common air, that it is so inert. In combination it always plays an active part. All substances containing it have a tendency to be decomposed. Chemically, that is, intimately united with oxygen, it forms one of the most violent agents known.

31. Oxygen combines with nitrogen in five different, perfectly definite proportions, by weight, viz.:

Protoxide (first oxide) of	nitrogen is	14	parts	of nitrogen	with	8	of oxygen.
Deutoxide (second oxide)	"	14	"	"	"	16	"
Tritoxide (third oxide)	"	14	"	"	"	24	"
Peroxide (highest oxide)	"	14	"	"	"	32	"
Nitric acid, aquafortis, is		14	"	"	"	40	"

It seems a very surprising and wonderful thing that these two gases should always unite in such exact proportions; that 14 parts by weight of nitrogen should always unite with exactly 8, or twice 8, or three or four times or five times 8 parts of oxygen. Yet this is always the case. And not only do nitrogen and oxygen unite in this exact manner, by this precise law, but all the other elements unite with each other in perfectly definite, invariable proportions. *How* this happens no one knows. All we can say is, that the Creator has made things in this manner, so as to unite according to this law. And this is called the **Law of Definite Proportions.** For when things always happen exactly in one way, we say that they happen *according to a law of nature*. It is inconceivable that they should always come so by accident.

This law is *universal.* Oxygen always unites in the proportion, by weight, of 8, or some multiple of 8. Nitrogen always in the proportion of 14; and *every* other *element* has its definite combining number. The combining number for hydrogen is 1; for carbon, 6; for sulphur, 16; for iron, 28.

And it is found that 9 pounds of water consist of 8 pounds of oxygen and 1 pound of hydrogen; and that 28 pounds of iron unite with 8 pounds of oxygen to form rust or oxide of iron. "Take, for example, 9 pounds of water, pass its steam over a known weight of pure iron turnings, heated red-hot in an earthen tube. No steam escapes from the tube, only air, which may be inflamed and burned. It is hydrogen gas, one of the constituents of water. That liquid has been *decomposed.* What has become of its oxygen? It has united with and oxidated the iron. What proportion of the water did it form? 8-9ths." If the iron be weighed, it will be found 8 pounds heavier. Subtracting from the 9 pounds of water, 8 of oxygen, the balance, 1, is hydrogen.*

If the experiment be very carefully conducted, it will be found that 28 pounds of iron have been converted into iron rust, and that all the rust formed by 8 pounds of oxygen weighs 36 pounds.

The several elements, or simple, uncompounded substances, are, for convenience, represented by the initial letters, and the proportions in which they unite by numbers placed a little above them. Chemists suppose that it is only the least possible, indivisible particles of matter or *atoms*, that unite, and that the atoms combine, 1 with 1, or 1 with 2, or with 3, or 2 with 3, and so on.

Oxygen is represented by O; Hydrogen by H; Nitrogen by N; Carbon by C; Sulphur by S. H O is water, because one atom of hydrogen is supposed to unite

* Dana's Muck Manual, p. 44. Whoever wants to get a vast deal of knowledge upon the subject of fertilizers, philosophically stated, in a small compass, may consult this valuable volume.

with one of oxygen. $N H^3$ or Am is ammonia,—three atoms of H and one atom of N. Carbonic acid is $C O^2$, that is, one atom of carbon with two of oxygen. N O is protoxide of nitrogen, one atom of each element; $N O^2$, $N O^3$, $N O^4$, $N O^5$, represent the successive oxides of Art. 31, and nitric acid, in which one atom of nitrogen is supposed to be united with five of oxygen; and if each atom of nitrogen weighs 14, each atom of oxygen must weigh 8, on the same scale.

32. **Nitric Acid,** like sulphuric acid, is so excessively *corrosive* as speedily to destroy almost any substance exposed to its action. It is a liquid, looking somewhat like water.

A flash of lightning, in the air, often causes oxygen and nitrogen to combine, forming nitric acid, which is immediately dissolved by the rain, and is sometimes found in rain water.

33. **Carbonic Acid** is the gas which rises, in the form of bubbles, in the fermentation of beer, or when you open a bottle of beer, or in the effervescence of cider or of wine. It is the gas which kills a person who remains too long in a close room where there is a pan of burning coals. It is formed by the combination of oxygen with carbon or charcoal.

34. All kinds of wood and other vegetable substances are made up mostly of carbon or charcoal, united with water, or with oxygen and hydrogen, in nearly the same proportions in which they form water. When wood is kindled, it unites with the oxygen of the air. Burning or combustion is the uniting of a combustible substance with oxygen, accompanied with light and heat.

35. The blaze or **Flame** is formed by the uniting of oxygen with a combustible gas.

36. **The Light** and **Heat** both come from the wood as it burns. While a tree is growing, it receives, from the sunshine, light and heat, and absorbs them, and lays them up in the wood. There they lie, as in a storehouse, till they are brought out by burning.

37. **Ammonia.** Hydrogen combines with nitrogen to form *ammonia*, which is one of the essential articles in the food of plants.

38. Wherever decay or decomposition of any animal substance, or almost any vegetable substance, takes place, there both these gases, *hydrogen* and *nitrogen*, are given out, and, at the very moment they leave the other substances with which they have been combined, they unite and form ammonia, which rises and floats in the air, and is dissolved rapidly by the moisture in the air, and is then brought down to the earth in the rain.

39. The little delicate roots absorb it from the earth, and it is carried into every part of the plant. Some power in the plant separates the two again, for both are always found in the growing parts; and nitrogen and hydrogen are found in the seeds.

40. Hydrogen unites also with *sulphur*, and forms a very offensive gas, called sulphuretted hydrogen; and this also enters into the composition of plants, as a part of their food.

41. In 100 pints of common air, perfectly *dry and pure*, there are about 21 of oxygen and 79 of nitrogen; that is, not far from one-fifth of oxygen and four-fifths of nitrogen. In its *common* state, 100 pints of air contain from 1 to 2½ pints of watery vapor; and 1,500 pints contain 1 pint of carbonic acid.

42. In breathing, the air enters into the lungs, and there the oxygen comes in contact with a portion of the

blood, and combines with it, much as oxygen combines with fuel in burning, and by this combustion sustains the animal heat, and keeps the body warm. When the air in the lungs is breathed out, it contains less oxygen than the air which had entered. In place of this oxygen which has staid in the body, a portion of carbonic acid is breathed out, which poisons, to a certain extent, the surrounding air. In this way the purity of the air would soon be destroyed, and it would be rendered unfit for breathing, if pure air were not brought in.

The quantity of air thus rendered unfit for *respiration* is known, and we can calculate exactly the space and the number of cubic feet of air which ought to be provided in chambers for men, and in stables and other places for other animals, according to the number and size of the animals to be shut up in them.

A Man needs from 200 to 350 cubic feet of pure air every hour. Supposing a person to require only 250 feet an hour, a close room of 10 feet in each dimension, having its air rendered more and more impure, by his breathing it, will, in four hours, be foul and very unwholesome, and wholly unfit to breathe.

43. It is thus plain that every place occupied by a living being, particularly by night, ought to be ventilated. That is, it ought to have a communication, by means of a chimney flue, or in some other way, with the pure, open air. Neither the body nor the mind of a person who has to breathe, night after night, the close, foul air of an ill-ventilated room, can remain healthy.

44. Plants do not breathe as animals do. But air is just as essential to them, penetrating freely into the tissues of their green portions, and there playing a part

necessary to their existence, and not wholly unlike animal *respiration.*

45. By daylight, and especially in the sunshine, plants absorb carbonic acid, turn the carbon, and water, or the elements of water, into the substance of the wood, stem, leaves and the other solid parts, and throw back part of the oxygen into the air. Growing plants are thus continually acting to purify the atmosphere, by taking up the carbonic acid which is poured into it by combustion, by decay, and by the breath of animals, and giving back oxygen suitable for healthy respiration.

We thus see the wise and beautiful **Relation** which has been established between animals and plants. The wind which blows from the habitations of men and animals carries foul air, no longer fit to be breathed, away to the woods and fields. There the plants extract from the air all that is poisonous; and the wind which blows from the field and forest brings back only the pure and vital element of oxygen, mixed with harmless nitrogen.

46. In the *night* time plants do not exercise this beneficent influence. On the contrary, they then exhale carbonic acid, at least in small quantities. It is this, perhaps, which renders it unsafe to have plants, especially when in flower, in a sleeping room.

It would seem that wood or woody fibre is not formed during the night, but that the presence of the sun's light is essentially necessary to this action of the life of a plant.

47. The oxides of the metals, and some other compounds, are *bases;* that is, they unite chemically with *carbonic acid*, *sulphuric acid*, *nitric acid*, and other acids, and form salts, called carbonates, sulphates, nitrates, and other *ates.*

They have been named **Salts**, from their resemblance to common table salt, though their properties are usually very different.

48. *Carbonic acid*, for example, intimately combined with lime, forms a salt called carbonate of lime, which is chalk or limestone. *Sulphuric acid*, combined with lime, forms sulphate of lime, or plaster of Paris. *Nitric acid*, chemically combined with potash, forms nitrate of potash, or saltpetre. All these are salts of great importance in agriculture.

49. Oxygen is also continually combining with wood and other vegetable substances. The decay of the fallen leaves is produced by oxygen slowly combining with the carbon of the leaves. Moisture and warmth are favorable to this combination, or oxidation, and heat is always produced by it. A heap of leaves, decaying, grows warm and continues warm till they are all turned into leaf mould, geine or *humus*. So the very gradual decay of trunks of old dead trees, and of every thing made of wood, is principally owing to the combination of oxygen with the carbon in the wood.

Nearly all decay is produced by oxygen. It is oxidation. During the process of decay of vegetable substances, not only carbonic acid, but, previously, humic acid, (from *humus*, earth,) and ulmic acid, (from *ulmus*, an elm,) are formed. Both these are made of carbon, hydrogen and oxygen, and both are elements of the food of plants.

50. **Humus**, or *geine*, in all its states, is a compound of carbon, with the elements of water, oxygen and hydrogen. When decay has just begun, the decaying substance is called *ulmin;* with a little more oxygen, it becomes *ulmic acid.* In both these, there is more hydrogen than is necessary to form, with the oxygen, water.

51. With the addition of more oxygen, just enough to form water, *humin* and then *humic acid* are formed. By the addition of still more oxygen, the humus is turned, successively, into geic acid, crenic acid, (*krenê*, a fountain,) and apocrenic acid. Several of these are often found, at once, in a mass of humus.

52. If nitrogen be present in a moist, decaying mass of substance, it unites with hydrogen, and forms ammonia; and a part of the ammonia, acted upon by oxygen, is gradually turned into nitric acid.

CHAPTER III.

THE ATMOSPHERE AND THE FORCES ACTING IN IT.

53. The air forms about the earth a coat which we call the **Atmosphere,** (vapor-ball,) and which extends upwards forty or fifty, perhaps two or three hundred, miles from the surface of the earth.

54. The atmosphere is the great ocean in which all animal and vegetable lives exist, and all the influences and agencies which act upon them are at play. Among these are *light*, by which all visible things are made known to us; *heat*, which pervades, and expands, and moves all things, and is essential to the life both of animals and of plants; *moisture*, alike essential, and by which nearly all things are softened or mollified; *sound*, without which the earth would be a silent desert, and voice and music and the pleasure of social life could not exist; and the wonderful cause of thunder and lightning, which we call *electricity*.

55. In the atmosphere, great operations are going on; all things are perpetually mingling, or trying to mingle. The winds are blowing, in vast circuits, from zone to zone, bearing heat from the equator and cold from the poles, moisture from oceans, lakes, and streams, and dryness from the mountains and plains, and scattering dust and the seeds of plants and the eggs of minute animals.

Into the atmosphere are continually *rising* vapors and exhalations from all moist and all decaying substances; poisonous *gases* from the breath of man and other animals, and from burning volcanoes and the fires which are kindled by accident, or for the uses of man. All these are constantly striving to diffuse themselves, and to penetrate and mingle with each other and with parts of the solid earth.

56. The *sun* is continually darting his rays of light and of heat in every direction, illuminating and warming every thing within the sphere of their influence. Every star, every fire, every candle is doing the same. *Oxygen* is always tending, with ceaseless effort, to enter into and combine with other things. Every other gas and vapor is, by its nature, diffusing itself in like manner. *Water* moistens, that is, enters into, every thing with which it can come in contact—the air, and all things in it, the earth, and the solid rocks.

57. And this it does by that force by which particles near each other are drawn nearer. It is this force which makes the particles of water rise upwards from the ground into a heap of ashes or fine sand, and penetrate among the fibres or grain of wood. It is this which draws water up into a tube of glass with a bore as fine as a hair, whence it is called **Capillary Attraction,** (from *capillus*, Latin, a *hair*.)

58. Another cause of the penetration of water is the force which draws fluids of different densities through a partition of thin skin or film placed between them, and makes them mix. This is called **Osmotic Action.**

We can easily conceive how this action takes place. Water spreads itself continually, and enters into whatever is in contact with it more readily than any other fluid. Thus it moistens and gets through a film more rapidly than the fluid on the other side, which also penetrates, but less readily. Both of them continue to move on, but the water always more rapidly.

59. *Oxygen* combines with the particles of metals and turns them into rusts or oxides; and, aided by moisture and warmth, it unites with the elements of wood and all other things made of carbon and hydrogen, and causes them to decay.

60. Do not the heavy gases, like carbonic acid, sink to the bottom of the atmosphere, and the light ones, like hydrogen and carburètted hydrogen, rise to the top?

No. *Each* gas spreads or diffuses itself throughout all the atmosphere. As much carbonic acid is found at the top of a mountain as in the bottom of a valley. If a plant has an attraction for ammonia, it draws to itself the ammonia near it, and combines with it; but the ammonia at a distance rushes in, comes near, and is attracted and combined also, and streams of it keep coming in from all quarters.

61. **Heat,** too, spreads itself, unceasingly, in every direction, and that in two ways. If it spreads from particle to particle, as it does in a piece of iron, or any other solid, or as it does in the earth, it is said to be *conducted,* or to spread by *conduction.* If it darts out, as it does, in straight lines, from all things surrounded by air or open

space, it is said to spread by *radiation*. As it spreads, it expands every thing; and as the temperature is every where continually changing, from winter to summer, from day to night, and every hour of the day and night, all solids must be constantly expanding and contracting, and the particles of which they are composed must be continually approaching to and receding from each other.

In *liquids*, the particles that are warmed expand and rise, while those that are cooled contract and sink, thus producing currents upwards and downwards in the liquid. *Particles* of other substances, floating or suspended in the liquid, as they become warmer, rise towards the surface, and, as they cool again, sink towards the bottom; or, if one side of a particle expands more rapidly than another, it turns over, seeming as if it had life and voluntary motion.

The vapors and *gases*, expanded by heat, become lighter, rise upwards towards the surface of the atmosphere, and their place is taken by cooler ones from every side.

62. Why does not this perpetual strife of forces produce disorder and chaos?

These forces are not lawless forces. They all have their limits within which they are compelled to abide. Besides, there are other mighty forces always acting against them, and constraining them to keep within their bounds.

63. One of these forces is the **Attraction of Gravitation**, which makes a stone fall to the ground, and draws every particle, every atom, towards every other, and all towards the centre of the earth, and the earth itself towards the sun. This gives them all their weight, and brings them to rest, and keeps them in their places.

Another is the force which binds the particles of a stone or of any other thing together, and makes it hard or strong or tough, which force we call the **Attraction of Cohesion.** Another is the force by which different things stick to each other, as mortar to a brick, or glue to wood, which we call the **Force of Adhesion.** And there are doubtless other forces which we do not so well understand.

64. One of these unknown forces is the **Force of Vegetable Life,** which draws into a growing plant the several substances which are necessary to its growth, and out of them forms all the parts of the plant. *Another* is the **Force of Animal Life,** which turns its food into the flesh and bones and other parts of the animal.

A third is the **Power which the Light of the Sun exerts** upon all vegetables and animals, upon all colors, perhaps upon all things within its reach.

A fourth is the power by which *electricity* draws light bodies, and perhaps heavy ones, towards an electrified surface, and again repels them.

65. It is from the influence of the sunlight that the carbonic acid and water in the sap of growing plants are turned into the substance called woody fibre, which gives them their hardness and strength. A woody plant, growing in the dark, lengthens, but forms no true wood, and so has no hardness.

It is the influence of this light which causes the *evaporation* at the surface of the leaves, which thickens the juices, and changes them into nourishing sap. Without the sunlight, the peculiar odors and tastes are not formed, nor all the beautiful variety of colors.

66. That the light of the sun has this great power over plants, is shown by the fact that most of those plants which naturally grow in places where the sunshine daily

comes, refuse to grow in the shade. Or, if one grows in the shade, it has none of the sensible properties, neither the strength, nor hardness, nor color, nor smell, nor taste, which it would have had growing in the sunshine.

In the growth of a tree, the stronger and fuller the light to which it is exposed, the greater the amount of carbon which is formed into its texture, and the harder and more compact its wood.

67. A single experiment shows that it is *light* and *not air* which gives wood its strength and hardness. Plant a little tree in a dusky room, with two openings, one admitting light but no air, the other air but no light, and all the little branches will soon turn towards the light.

68. This seems to be because on the side of a branch towards the light, *wood* is formed, the growth is checked, and the branch hardened ; on the other side, growth continues more rapidly, and the parts lengthen, and thus bend the little branch over towards the harder side.

During very warm, moist *nights*, plants may grow in length and in every other dimension. In the sunlight only do they form wood. Hence it is that in seasons of unusual sunshine, the wood in a tree fully exposed to the sun is formed with more than common perfection, as is also the bark.

69. The power of the sun's light upon animals is not less striking. The animals,—beasts, birds, fishes and insects,—of the torrid zone, where light is intense, have more activity, more vivacity, and more brilliant colors than animals of the temperate and frozen zones. All animals suffer from being shut up away from the light.

70. **Human Beings,** not less than other animals, **Suffer from being kept away from Sunshine.**

A child properly managed, and left to spend a good many hours every day in sunshine, has more color, more strength, more activity, more health, and better spirits, in consequence. A child kept away from the sunlight is pale, weak, dull, delicate, and sad, and is liable, when this exclusion from the sun's light is long continued, to many forms of fearful disease.

71. The sun, and, with it, the air, are constantly acting, with great power, upon the soil.

The heat of the sun swells or expands the particles, and thus makes room for the entrance of the air; and the oxygen of the air and the other gases which float in the air combine with some of the elements of the soil, and render them fit to aid in the growth of plants. Other beneficial effects are produced, of which more will be said hereafter. All these are increased by the frequent stirring of the soil.

Hence it is that when trees are to be planted, it is important to dig the holes some time beforehand, in order that the fresh earth in the holes may be acted upon by the sun and the air as long as possible.

72. The atmosphere produces many other different effects upon animals, upon plants, and upon the soil, varying with the direction and force of the winds, heat and cold, the weight and the moisture of the air, rains and droughts, dews, clouds, and fogs, mists and storms.

73.* What is **Electricity**? We know it only by its effects. If we rub a rod of amber, or sealing wax, with a piece of woollen cloth, the amber or wax is immediately excited, and draws towards itself, or *attracts*, light bodies, such as bits of thread, or of elder-pith hung to a thread. The cause of this attraction was called *electricity*, from its being first observed in excited amber, which the Greeks called *electron.*

A rod of *glass* may *be excited* in the same manner by rubbing with silk. But in this case the electricity is of a different kind.

74. Take a smooth piece of iron or brass, or any other metal, and hang it up by silk threads so that it shall not touch or be near to any thing, and fasten to it several pith balls hung to the end of cotton threads. Rub the piece of metal with a rod of excited amber or sealing wax, and, immediately, electricity is excited, and the pith balls are repelled, and fly from each other and from the metal as far as they can go. Bring the rod of amber or wax near to the balls, and they will be repelled and avoid it. But if you bring a rod of excited glass near them, they will be attracted, and will fly towards it. The electricity excited in the glass is of an opposite kind to that excited in the amber, and the **opposite electricities attract each other.**

75. Touch the metal with a finger, and the little balls immediately fall together again. The electricity is *discharged* through the finger.

76. Something similar is supposed to take place with vapor. When water is turned into vapor by the sun's heat, it forms little hollow bubbles or vesicles, which repel each other in consequence of being electrified by evaporation. Any thing which draws off the electricity of a cloud of such vapor causes the little vesicles to collapse, and rush together, and form drops of rain.

CHAPTER IV.

CHANGES IN THE ATMOSPHERE.—INSTRUMENTS TO MEASURE THEM.—CLIMATE.

77. The state of the atmosphere is continually changing, and several instruments have been contrived for the purpose of measuring its changes, and of showing what its state is. The three most important are,—

(1.) The *thermometer*, (heat-measurer,) which shows the changes in the heat of the air;

(2.) The *barometer*, (weight-measurer,) which shows the changes in the weight or pressure of the air; and,

(3.) The *hygrometer*, (moisture-measurer,) which shows the changes in the amount of moisture in the air.

78. The **Thermometer** is constructed on the principle that almost every substance known is swelled or expanded by being heated, and contracted by becoming cooler; and that the expansion is in proportion to the degree of heat.

79. This may be proved by various experiments. If a hole in a plate of iron is just large enough to admit a rod of iron when cold, it will be found that, when the rod is heated, it will no longer enter. If the rod be left to cool down to its former temperature, it will enter as at first. This shows that the rod has been expanded by heat, so as to take up more room than it had previously taken up.

When a wheelwright makes an iron tire for a wagon wheel, he makes it just long enough to bring the fellies closely together. In order to do this most effectually, he makes it a little too short to go on while cold. He therefore expands it by placing it on a circular fire, and when it is hot, he easily slips it on. Upon cooling, it contracts, and so draws the fellies firmly and closely together.

80. There are several kinds of thermometer. That in common use in this country is called *Fahrenheit's*, from the name of the person who first made it. It is made of a glass tube (A B fig. 1,) having a small bore, with a bulb (A) at one end, filled with quicksilver, and fastened upon a plate of metal or other substance, which is to be marked with degrees. When it is to be marked, or *graduated*, the bulb and tube are held in a mixture of melting snow, or of snow or ice and water. The quicksilver within the tube contracts and falls to a certain point, where it remains. Just against this point a line is drawn on the plate of the frame, and the number 32° (thirty-two degrees) is marked at the end of it. This is called the *freezing point.*

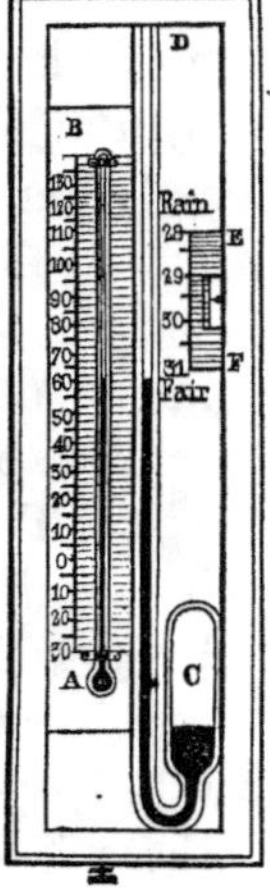

Fig. 1.

The thermometer is then held in boiling water. The quicksilver expands and rises till it reaches a point at which it remains stationary. Against this point a mark is drawn on the plate, and the number 212° (two hundred and twelve degrees) is made. This is called the *boiling point.* The space between the freezing and boiling points is divided into 180 equal parts, called *degrees.*

The space below the freezing point is divided into parts of this same length, down to the bulb. A thermometer, to be used to measure extremes of cold and heat, should be long enough to extend from 40° or 50° below the freezing, to a few degrees above the boiling point. But as this requires a long tube, instruments for common use are marked up to only 140° or 150°.

The thermometer is an instrument of great utility to the farmer, and indeed to every intelligent person.

81. A **Barometer** is constructed upon the principle, now a well-known fact, that air has weight. It can be weighed by a delicate balance, by first filling a flask with air and weighing it, and then drawing out the air by an instrument called an air pump, and weighing the flask without the air. At the level of the sea, one hundred cubic inches of air weigh 305 grains, while water weighs 816 times as much.

82. The air seems to be pressed towards the earth by its weight, just as water is kept in the ocean and in lakes by its weight. Its pressure is greatest at the level of the sea, because of all the air in the sky above. As we ascend a hill or mountain, the pressure becomes less, because there is less air above us, and because the attraction of gravitation is diminished. The air is constantly in motion; and its pressure upon the surface of water, and upon all other surfaces, is constantly varying. The purpose of a barometer is to measure this varying pressure.

83. A barometer is made of a large tube of glass, precisely like that of the thermometer (A B fig. 1,) but much longer,—not less than 32 or 33 inches long,—with a bag or bulb at one end, filled with mercury, or quicksilver, so contrived as to rise to a certain height in the tube, while it has the air bearing upon it in the bag or bulb. From the upper end of the tube the air is first completely withdrawn or exhausted, by the tube's being held upside down. The tube is then turned back and fastened to a wooden frame, or enclosed in a case with a graduated plate behind the upper end of the tube, on which plate are marked the heights of the column of quicksilver.

84. There is no pressure upon the top of the mercury in the tube, and the pressure of the air upon the mercury in the bag forces up the mercury in the tube till its weight exactly balances the weight, or downward pressure, of the air. The pressure of the air is sometimes greater, sometimes less, but is commonly sufficient to balance the downward pressure of a column of mercury 29 or 30 inches long. As the pressure of the air increases, it causes the mercury to rise higher; as it diminishes, it allows the mercury to fall lower; and these changes are seen, by observing how high the surface stands as marked on the graduated scale. Changes in the weather are sometimes foreshown by changes in the height of the mercury as indicated by this scale.

85. The downward pressure, or weight, of a column of mercury 30 inches long, and an inch square at the bottom, is 15 pounds; and as this column is sustained by the pressure of the air, every where near the level of the sea, we conclude that the pressure of the air, on every square inch, is 15 pounds.

86. When the mercury in the tube is slowly and gradually rising, it commonly indicates the approach of fine weather. When it is regularly and slowly falling, it indicates foul weather. A rapid and sudden fall of the mercury threatens a violent wind.

While it is rising, the surface of the mercury is convex, or swelling upwards; when falling, concave, or hollowing.

87. A very compact and convenient barometer is made at Lowell, Mass., of a somewhat different construction. A short column of mercury, in a glass tube, (C D fig. 1,) is pressed upon, at the upper surface, by the atmosphere, with which it has communication. The other end of the

column of mercury presses upwards upon perfectly dry air *confined* in an enlargement (C) of the bent tube.

When the weight of the atmosphere increases, the mercury is pressed *downwards* in the long arm and rises in the short arm of the tube, the dry confined air, from its elasticity, yielding to the pressure. The length of the column of mercury is marked upon a graduated scale placed on one side. A movable scale, (E F) called a *vernier*, is attached, so contrived as to measure the height of the column to hundredths of an inch.

88. Careful observation of the winds, and of the barometer, with a knowledge how to observe, will often enable a person to foresee rain for some hours, or a day, or possibly longer, before it comes; but no person can yet predict, with any certainty, whether the succeeding month will be dry or rainy.

It is only of late that careful and continued observations have been carried on, upon a large scale, to discover the *laws* of storms. It is found that nearly all storms, in the Atlantic States, come from the west, and travel pretty rapidly from west to east. Hereafter we may know, certainly, the approach of a storm many hours before it reaches us. Prof. Henry, at the Smithsonian Institution in Washington, having telegraphic communication with many parts of the country, is usually able to predict the approach of a rain-storm twelve hours before it comes.

89. Of what are commonly considered the **Signs of Rain,** none are entirely reliable. When the sun sets clear, with a westerly wind, and the clouds float high and in round, compact, well-defined masses, we may expect the next day to be fair. But when the sun sets in a deep mass of

cloud, with a southerly wind, rain may be expected, that night or next day.

When the swallows fly low and often dip their wings in the water over which they are flying, when the crow cries louder and more frequently than common, when water-fowl are very noisy and active, when dogs appear unusually dull and sleepy, when pigs run about and look uneasy, when the croaking of frogs is loud and general, when earth worms are seen in great numbers on the surface, some people expect rain.

90. The principle upon which the **Hygrometer** is constructed is the fact that there is always more or less moisture in the air, and that this moisture is absorbed by certain substances, making them heavier, and enters into lines or cords made of other substances, making them thicker and shorter.

91. A hygrometer may be made of a piece of sponge filled with a solution of some salt, which has an attraction for water. This sponge is suspended to one end of a balance, and, as it grows heavier by the moisture absorbed, causes the other end to rise, and thus indicates the quantity of moisture in the atmosphere. Or it may be made of a cord or string, with a weight attached, placed over a pully, and showing the moisture by its lengthening or shortening.

92. A still more delicate hygrometer is formed of two thermometers on the same frame, the bulb of one of which is covered with thin gauze, which may be kept continually moist by a contrivance like a wick, communicating with a cylinder kept full of water. The moisture on the gauze evaporates and cools the bulb within. The amount of evaporation depends upon the dryness of the

atmosphere, and is shown by the difference between the two thermometers.

93. By means of these three instruments, and knowing how to use them, an intelligent husbandman may select the moment most favorable or most important for certain operations; and can often predict, with an approach to probability, what changes will take place in the weather before night or before the next morning.

94. The **Variations in the Temperature** of the air depend first, upon the seasons,—from the cold of winter to the heat of summer; 2d, upon the direction of the wind,—some winds always bringing cold, others always bringing heat; 3d, upon the clouds, which prevent the sun's light and heat from falling upon the earth.

95. The atmosphere being in continual motion, like the waters of the ocean, the column of air over us is sometimes longer and heavier, and sometimes shorter.

96. **Variation in the Moisture** of the air depends chiefly upon the winds, which bring on air more or less abundantly charged with moisture, according as they have passed over seas, lakes, or rivers, or over a continent. In the Atlantic States of America, the easterly and southerly winds, coming from over the ocean, are always full of moisture. The south and west winds, coming from warmer regions, are warm, and, in proportion as they are more westerly, are dryer winds. The north and west winds, coming from the mountains and plains of the continent, are dry and cold. The coldest and dryest are the north wind and the north-west wind, and any wind from a point between the two.

The moisture also depends on the temperature. Heat dissolves moisture as water dissolves salt. When the air is warm, it can contain a great deal of moisture; but as the air cools, the moisture in it is condensed into clouds, fogs, or mists, and finally into rain.

97. There are many other *atmospheric appearances or phenomena* which it is important for the husbandman to be acquainted with, such as dew and hoar frosts, which take place during the night, when the sky is clear; snow, which seems to be frozen mist; hail, and hurricanes, which are by some persons attributed to the action of electricity.

98. The **Formation of Dew** depends upon a property which all solid substances have, in a greater or less degree, according to their nature and outer surface.

When I hold my hand towards the fire, I feel the heat darting out from the fire to my hand. I feel it darting out, in the same manner, from a hot stove or from a hot flat-iron, on whatever side of the stove or iron I hold my hand. The heat which darts out thus in every direction from any hot thing is said to *radiate* from it, because it comes out straight from it, just as the spokes, (*radii*, in Latin,) come out on every side from the hub of a wheel. If I observe carefully, I find that the heat comes out more abundantly from a stove the surface of which is very rough, than from one which is very smooth; and I discover that the reason is, that every little projecting point radiates a stream of heat.

Now, what I find to be true of the surface of a hot stove is true of every surface. Every solid body is continually sending out heat in straight lines,—radiating heat,—from its surface. If several bodies are heated to the same degree, the one which is roughest will radiate and consequently cool most rapidly.

When the sun sets, all things which have been exposed to his heat send it forth by radiation, and grow cool.

Those things which have the roughest surface, like the stems and leaves of grass, cool most rapidly. The heat thus radiated is sent out into the thin air, and, if there are no clouds, is lost in vast space. The air which is near to these blades of grass imparts its heat to them and grows cold. The air thus becomes incapable of holding in solution all the water it had dissolved, and deposits it, in minute particles, upon the surface of the grass. The radiation goes on, and the moisture continues to be deposited, till the blades of grass are covered with drops; and these drops are drops of dew.

Now, just as, by placing a screen before a fire, we prevent the heat from being radiated into the room, and send it back to the fire, so a screen of clouds stretched over the earth prevents the heat received from the sun from being rapidly radiated into the empty air, and thus prevents the formation of dew. We find, accordingly, that dew is formed only on clear evenings.

99. **Hoar-frost** is formed in precisely the same manner as dew, but at so low a temperature that the moisture freezes as it collects on the radiating surface, and, instead of forming round drops, shapes itself into slender needles of ice.

100. **The Climate of a Country** is the general effect of the combined action of all the causes just spoken of, viz., heat, moisture, wind, and of others still.* The husband-

* Humboldt says: "The expression 'climate' signifies all those states and changes of the atmosphere which sensibly affect our organs—temperature, humidity, variation of barometric pressure, a calm state of the air or the effects of different winds, the amount of electric tension, the purity of the atmosphere or its admixture with more or less deleterious exhalations, and, lastly, the degree of habitual transparency of the air and serenity of the sky, which has an important influence not only on the organic development of plants and the ripening of fruits, but also on the feelings and the whole mental disposition of man."—*Cosmos*, I. 313.

man ought to understand the climate of the country in which he lives, in order that he may accommodate himself to it in the management of himself and of the animals and plants he has charge of.

101. Our New England climate is one of extremes. The heat is very great in summer, and the cold very severe in winter. The climate of the west of Europe is far milder. As we go west from the Atlantic the climate becomes less extreme.

102. So great is the influence of climate that each country has its own peculiar productions, which it is often difficult to *acclimatize*, that is, make to flourish, in any other; and, before introducing a new plant or animal upon his farm, the farmer ought to ascertain whether it is suited to the climate. But both plants and animals from distant countries are frequently introduced with success; so that, without a fair trial made by himself or some one else, the farmer ought not to take it for granted that a new plant or a new animal will not be safely and successfully introduced.

103. The **Diversity of Climate** depends on many causes; some general and some particular and local. Among the general causes, the first is latitude, or the distance from that part of the earth where the sun is at noon directly, or vertically, overhead. The heat depends, in a great measure, upon the height above the horizon to which the sun rises at noon. The higher it rises, the hotter it is.

The second cause is elevation above the level of the sea. The higher we go above this level, the colder we find it, till we reach the tops of lofty mountains, where the snow never melts.

The third cause is distance from the sea. Nearness to the sea has a tendency to moderate the cold of winter and the heat of summer; and islands in the ocean have usually a more equable climate than any part of a continent.

Another cause, particularly affecting the ripening of fruits, is the brightness of the sun, from the clearness of the atmosphere. The heat of clear, uninterrupted sunshine ripens fruit more rapidly and develops the sweet and rich juices more effectually than the same amount of heat under a cloudy sky.

104. Some of the particular and local causes are the condition of the surface of a country, whether it is covered with woods, or bare, situated on the mountains, on a plain, on the side of a river, or at the bottom of a valley, protected against the prevailing cold or hot winds, or exposed to them; and the nature of the soil, its inclination, and its exposure to the south or north, to much or to little sunshine.

CHAPTER V.

OF WATER.

105. Though it seems so simple and pure, yet water is, as has already been said, a compound of the two elementary substances, oxygen and hydrogen. As it is of vital importance, in the economy of nature, it is found in the greatest abundance, filling lakes and seas and oceans.

It is indispensable to the nourishment both of plants and of animals; and it dissolves much of the other food with which plants are nourished.

106. At the usual temperature of the grêater part of the year, water is a transparent liquid, which, when pure, has neither color, taste, nor smell. But while water is the great solvent of vegetable food, it is itself dissolved by heat, a still more powerful solvent.

107. Water is found in the three *forms* or conditions of ice, water, and vapor, according to the amount of heat with which it is combined.

(1.) With little or no heat, it is solid **Ice** or snow. If extremely cold ice be placed in a kettle over a fire, it will be found, by observing a thermometer with its bulb placed within it, to rise gradually until it reaches 32°. It then begins to thaw or turn into water, and if a steady fire be kept up, under the kettle, it continues to thaw until all the ice becomes water. During all this time, though heat from the fire is constantly entering it, through the kettle, it continues of the same temperature, just at 32°.

What has become of the heat? It has been used up in dissolving the ice and turning it into water. It has not rendered the water warmer; it is hidden or *latent* in the water; and is called the **Latent Heat** of the water. Ice has been changed by combining with heat, into

(2.) **Water.** If, now, the same steady fire be continued under the kettle, the temperature of the water gradually rises to the boiling point, 212°, and then begins to boil. With the same steady fire, the water will entirely boil away, or *evaporate*, in a certain space of time. And it will be found that it takes more than five times as much heat to boil the water all away, as it had taken to raise it 180°, from the freezing to the boiling point. At the same rate, the water would have been raised nearly to 1,000°, if it had not been dissolved by heat and turned into

(3.) **Vapor.** The vapor thus formed is no hotter than the boiling water. It does not rise above 212°. What has become of all the heat? It has been used up in turning the water into vapor. This heat is not indicated by the thermometer. It seems to be latent in the vapor; and it is called the **Latent Heat of** the **Vapor.**

108. The *boiling* of water is the agitation produced by the rising of the vapor, formed at the bottom of the kettle, up through the rest of the water; and the vapor is more abundantly formed in proportion as the heat of the fire is greater. But the water does not change its temperature in consequence of the violent ebullition. For common cooking, therefore, the gentlest boiling is just as effectual as the most violent.

109. At the boiling point, vapor is formed very rapidly. But water, exposed to the air, is continually evaporating, at every temperature. Indeed, such is the tendency of water to take the form of vapor, that even snow and ice, in the air, are constantly turning into vapor. Wherever it takes place, evaporation always uses up heat, or causes it to become latent, and thus cools the air and all surrounding objects. Indeed, whenever vapor, or air, or any other gas, expands, so as to occupy more space, it at the same time requires more heat and absorbs it from every thing within its reach capable of furnishing it. Its capacity for heat is said to be increased.

110. When, on the contrary, vapor turns again to the state of water, it *gives out* all the *latent heat* which it had taken in, while turning from water into vapor. The same is true of other gases. Whenever they are con-

densed, they give out the heat which had sustained them in the form of gas.

And, in like manner, **when Water Freezes, it gives out** the **Heat** which it had taken in, while turning from ice into water. We thus see why it happens that, to protect vegetables, in a cellar, against freezing, we have only to place tubs of water there, the warmer the better. The temperature of the cellar will not fall below the freezing point, till the water has been converted into ice.

111. The atmosphere always contains moisture; that is, water in the state of vapor, which the heat of the sun has drawn up from the surface of the earth and sea, and which floats, invisible, in the air. The warmer the air is the more water it can contain. When the air cools, the invisible vapor which it contained becomes visible in little hollow globules or vesicles, like minute soap bubbles, and forms clouds, fogs and mists.

112. The difference between clouds and fogs or mists is chiefly their situation. **Clouds** are at a distance or high up in the air; **Fogs** are clouds near the earth; and if the fog be thick enough to wet us considerably, without drops, we call it **Mist.** When a person, looking at a distant mountain, sees it capped with a cloud, another person, standing on the top of the mountain, finds himself surrounded by fog or mist.

113. **Rain.** The air itself may be capable of dissolving water, but the quantity which the air can hold depends upon its warmth.

Wind which has long been blowing over the sea becomes completely saturated with moisture in the state of vapor. If it now blow upon low land warmer than itself, the air becomes warmer and retains all its moisture; if upon land colder and gradually or rapidly higher, it is cooled and parts with its moisture. The vesicles of vapor are brought near each other, come together, and form drops large and heavy enough to fall, and which come down as rain.

If air full of moisture be met by air much colder than itself, the sudden cooling causes the water to be thrown down, or precipitated, in torrents of rain.

114. The cause of the fall of rain in a thunder shower is thought to be the fact that electricity is always evolved during evaporation, and that a cloud formed by evaporation must be therefore charged full of electricity. When a cloud so charged meets another, or a mass of air, charged with the other kind of electricity, the opposite electricities rush together and unite in a lightning flash, and the moisture held suspended by the action of electricity is precipitated to the ground.

115. When, during the formation of the rain drops, the temperature of the air is below the freezing point, the vesicles of moisture, or their fragments, are frozen into little icy needles, which unite, at an angle of 60°, into beautiful, star-like flakes of **Snow**, and fall to the ground.

Snow has been called "the poor man's manure." It always brings down with it fertilizing substances; and it performs a most important office in many regions, by covering over and protecting from extreme cold the surface of the earth with all its clothing of plants, and keeping in the warmth which had entered the earth during the previous summer, and preventing its being radiated away into empty space.

116. How **Hail** is formed is not perfectly well known. Hail seems to be drops of rain frozen. *Electricity* has something to do with it, and in some parts of Europe,

hail storms have been rendered much less frequent by the use of lightning rods.

117. **Springs.** The water which falls upon the earth in rain, sinks into the ground and moistens it; and, when very abundant, penetrates deeper, till it meets with beds of rock, or clay, or of some other impermeable earth, that is, earth through which it cannot pass. It runs along the surface of these beds until it meets a natural opening, out of which it issues as a fountain or *spring*. Or, it may remain in a basin, on the surface of the impermeable bed, and be safe, as in a reservoir, until an artificial outlet is made by digging a well.

118. From springs run little rivulets, by the union of many of which are formed brooks, rivers and lakes; the waters of all of which commonly flow at last into the sea. There, the heat of the sun raises it in vapor to begin again the beneficent circuit, and form mists and clouds and rain.

119. Water is essential to the life of every plant. Several of the substances on which plants feed, can penetrate into their cells and thence through the tissues, only after being dissolved in water. With it they are sucked in by the roots, and in it are carried to the very extremities of the plant.

120. Next to heat, water is the most universal solvent. The rain, as it descends, absorbs and condenses the gases which float in the atmosphere, and brings them down into the earth fit for the use of plants. Of *ammonia* it can dissolve 780 times its own bulk; of *carbonic acid*, its own bulk; and it commonly brings down a portion of air, rich in *oxygen*, and sometimes *nitric* acid. It also absorbs and brings down all kinds of dirt, and other impurities, numerous minute seeds of plants, and

invisible eggs of microscopic animals, and thus cleanses and sweetens the atmosphere.

121. Evaporation from the surface of the earth always cools it. But, on the condensation of ammonia, and the other gases, the reverse must take place. The heat which had held them in a gaseous form, is given to the water in which they are absorbed and condensed, warms it, and, sinking into the earth, warms the soil.

122. Plants absorb a large quantity of water through every part of their surface, but chiefly through their roots. But by the action of light and heat, they exhale a good deal of it through the leaves. You have only to cover a plant exposed to the sun's light with a bell glass, and you will presently see the inner surface of the glass covered with dew, and soon after with little drops. The evaporation which is going on from the surface of leaves is one of the sources from which the moisture of the atmosphere is supplied. As we are subject sometimes to excessive heat and drought, and sometimes to excessive rains, the object of the farmer should be to guard against both, and to render his fields, as far as he can, independent of variations in moisture.

123. We manage to prevent plants from suffering for want of water by irrigation, that is, watering with little streams, when these are possible and not too expensive; and by other artificial means. We can do something towards it, often we can do a great deal, by keeping the tops of the hills in our neighborhood covered with trees. These attract and impede the clouds, and induce them to pour down their rain.

124. Deep ploughing, by rendering the earth to a considerable depth capable of retaining moisture, will also

do something; and fertilizing with substances which attract moisture, will do still more.

Every thing done to improve the soil makes it retentive of moisture. Clay, mixed with a sandy soil, converts it into a retentive loam. The remains of vegetable and animal substances form a spongy matter in the soil, which acts as a reservoir to retain the moisture and other food of plants, and yield it only to their roots.

125. The rain, as it falls, always contains carbonic acid, ammonia, and other elements of plant nourishment. If it sink into the earth, the soil absorbs all these precious materials, and allows the superfluous water to escape only after having left its contribution in the soil. Besides, if the rain be allowed to run off from the surface, it forms streams and little torrents, and carries with it much of the loose and most valuable portions of the soil.

The soil should therefore be kept, for some depth below the surface, so mellow and penetrable, that the rain, instead of running off, shall sink into the ground. In ploughing a side hill, the furrows must run horizontally along the slope, so that each furrow may detain the water as it falls, and prevent its forming gullies, which it will do, if the furrows run up and down the hill.

126. Excess of wet is also sometimes to be feared, especially when the water has no way of running off, but remains stagnant, either beneath or above the surface, for it then causes the plants with which it comes in contact to mould and decay. We must then have recourse to ditching and drainage.

127. **Drainage** is an operation by which we draw off the superabundant water from the soil and from the earth lying beneath the soil, where it would not otherwise escape. It is effected by placing lines of porous earthen

tubes at a convenient depth, so arranged as to receive the superfluous moisture and carry it off.

128. The effects of drainage may be explained by a comparison. Plants which are kept in flower-pots would soon rot at the root, if the water with which they are watered were left to stagnate in the bottom of the pot without any means of escape. For this reason, the bottom of the pot has a hole in it, to let the superfluous water run out. Now drainage does the same service for the field that the hole in the bottom does for the earth in the flower-pot.

129. Drainage produces several other effects, three of which are important.

(1.) The earth being rendered less moist at the surface, far less evaporation takes place there. Whence, as evaporation always cools the surface very considerably, a drained field keeps in the heat better than one not drained; and the natural consequence is that the crops ripen earlier. The grain on a drained field is generally fit for the sickle some days, often some weeks, earlier than that on other fields.

(2.) Lands well drained and deeply tilled bear the drought better than others. The reason of this seems to be, that the pores are always open in deeply tilled, well-drained land, to an unusual depth. Evaporation cannot reach to a great depth, and, in a season of drought, the open pores allow the moisture which has been kept in the deep earth to rise by *capillary attraction.*

(3.) The subterranean pipes laid in the earth, open the soil to a freer access of air, allowing it, as it were, to breathe, and receive the benefits of being subjected to the action of the air. The soil is thus rendered fit to absorb and retain the nutritious substances brought into it by the rain water, and keep them laid up for the nourishment of plants.

130. Here then are the **advantages of** deep and **thorough drainage.** It deepens the available soil, by removing any superfluous water from the lower portion, and allowing the roots of plants to penetrate freely. It warms the land by diminishing evaporation at the surface. By carrying the redundant moisture readily away at all seasons, it gives the opportunity of early cultivation, thus lengthening our short seasons, and of thoroughly mellowing the soil, which cannot be done if it be too wet; and it entirely avoids the danger of losing the plants on the surface by having them freeze out, as they often do, if water continues to stand on the surface at the approach of very cold weather. It moreover guards plants against the evil consequences of drought.

131. For, in a well-drained soil, the roots will penetrate to a much greater depth than in an ill-drained soil. By draining, only the unnecessary and hurtful moisture is carried away. The soil, if rich, retains very tenaciously all that is necessary, and parts with it very reluctantly and only to the roots of plants. Now roots which have penetrated two or three feet have twice or thrice as large a store of moisture to draw upon, in case of drought, as those which have been prevented from going down more than one foot.

In a well drained field, the spring rains, instead of being allowed to run away and be lost, are saved, as in a reservoir, against the heats and drought of summer.

132. A rich soil, rendered deep and mellow by thorough cultivation, and by a system of underdraining, is thus the best preventive to the consequences of drought which the farmer can provide, and it is, at the same time, most effectual against the evils of excessive rain.

CHAPTER VI.

OF PLANTS.

133. Though fixed, and incapable of voluntary motion, and differing from animals in structure and organization, plants proceed from other parent plants, and live, are nourished and die, like animals, and, like them, produce offspring similar to themselves. Plants live and grow. Animals live, grow and feel. Vegetable life, therefore, is a very different thing from animal life.

134. The simplest of all plants consist of mere bladders or little round cells. These little cells imbibe their nourishment, in a fluid state, directly through the thin coat by which they are covered. The fluid within moves around in little curves, and changes at last take place in it, by which other smaller cells are formed. These gradually enlarge and finally burst the covering of the original cell, and become new plants, similar to their mother cell, and grow to the same size. Such are the simplest of all plants; and the growth of other plants, even of the highest perfection of structure, takes place by the formation, within the cells already existing, or outside of them, of other cells similar in nature but sometimes differing in shape.

135. Plants, consisting each of a single cell, are found in such numbers as sometimes to give a brilliant red color to whole miles of snow and ice on which they grow.

136. Other plants, almost as simple, are formed of a thread of single cells, strung together, end to end, like a string of beads. Of this structure are many delicate fresh water plants. And it is a plant of this kind which,

by growing very rapidly through dough, in which its seeds have been sown in the form of yeast, causes an action which makes it swell and form light bread. Other plants are formed of a single thickness of cells arranged side by side and end to end. These also are usually found growing in water.

There are still others which consist of a few, often only three or six layers of cells, plants having length and breadth with but little thickness. Such are the *lichens* which form a thin crust on the bark of trees and on the surface of rocks which have been long exposed to the atmosphere.

137. Most plants are formed of cells growing out of each other in every direction, upwards, forming the stem, downwards, forming the root, and on every side, forming the thickness of root, stem and branches, and leaves and flowers and fruits.

138. The parts just enumerated, the parts of which the plant is made up, are called the **Organs.**

139. The principal organs are 1st, the root; 2d, the stem; 3d, the leaves; 4th, the flower; 5th, the fruit.

140. **The Root** is the part which penetrates from the light into the earth, and gives the plant foothold, and the means of obtaining nourishment. It usually divides into smaller and smaller roots and rootlets, or radicles and fibres, more and more slender, the cells along the sides and extremity of which are the real mouths by which most of the food of the plant enters into its circulation. The amount of food which a plant can receive from the soil depends upon the number and surface of the fibres of the roots.

141. The **Stem** is the part of the plant which rises upwards into the air and light, and supports the branches,

leaves, flowers and fruit. The point at or near the surface of the earth, where the root and stem join, is called the *collar* of the plant.

142. The stem and branches are protected from heat and cold by the *bark.*

143. The **Leaves** are the organs through which the air, and the light and heat of the sun act upon the sap which comes up into them through the stem. Through their surface the superfluous moisture is evaporated, and oxygen gas is thrown out into the air, and carbonic acid and other gases for the nourishment of the plant are absorbed.

The **Sap** changed by these actions of the elements, is carried back down into the stem, and converted, by the vital action of the plant, into wood, bark, new branches and leaves, fruits and whatever else is produced by the plant.

144. **The Flower** is the organ by means of which the seeds are prepared; and a great object of the plant is the production of fruit containing seeds.

145. By carefully examining a rose, you may see the several **parts of which a Flower consists.** Outside of the flower leaves is a flower cup or **Calyx,** of five green leaves, called the calyx leaves or *sepals*, which cover and protect all the parts of the flower, before they are ready to open.

146. Inside the calyx are the flower leaves, called **Petals,** tender, and of a delicate texture and beautiful color. All the petals together are called the **Corolla.**

147. Next inside the corolla are the **Stamens,** slender threads or filaments, of a pale yellow color, each bearing at its extremity a little sack called an **Anther,** full of fine dust called pollen. This dust or **Pollen** is essential to the fecundation of the seeds, that is, to their becoming perfect, fertile seeds, fit to produce a plant.

148. Inside the stamens, in the middle of the flower, are the **Pistils,** each one of which consists of a short column, called a **Style,** tipped with a very delicate crest called the **Stigma,** which is usually tender and moist when the flower is in perfection. In a rose the style seems to be nearly wanting, the stigma appearing to rest almost directly upon the *receptacle* or centre of the flower. But if you cut down directly through the centre of the flower, you you will find the style somewhat long and connected at the bottom with an ovule.

149. The **Pollen** or fertilizing dust of the anther falls upon the moist stigma, and penetrates, by means of something which looks like a root, to the interior of the base of the style to a cavity called the **Ovary,** containing *ovules*, or imperfect, rudimentary seeds. The effect is to *fertilize* the ovules and make them become real, proper seeds, by producing within them an **Embryo,** or minute, future plant.

150. When the seeds are fertilized, the flower begins to fade. Its corolla falls off, its stamens shrivel up, and its calyx usually, but not always, falls or shrinks and disappears. The ovary swells and becomes the **Fruit,** which, in process of time, ripens and falls or dries up or decays, according to the kind of plant, and leaves the seeds ready to *germinate* or sprout, and thus become plants, or to be gathered and sown at the proper season.

151. Whatever contains the seed is properly called the **Fruit** of a plant. In the case of wheat, rye and some other seeds, each kernel is at the same time a seed and a fruit. Usually, however, a fruit contains several or even a large number of seeds. A bean pod or pea pod or a poppy head, is a fruit, as well as an apple, a pear or a melon.

We may now understand what is meant by organic substances. Plants, as we have just seen, are made up of organs. So are animals. The lungs are the organs of breathing, the stomach is the organ of digestion. All the parts of animals and plants are organized, and the substances which belong or have belonged to animals or plants are called *organic.* Mineral and all other substances are *inorganic.*

152. Now observe what happens when the seed is put into the ground. Every seed contains an *embryo* or minute plant. This, called the sprout, you may easily see in a bean, if you open it carefully. When a seed is put into the earth, in a favorable state of moisture and warmth, it presently begins to sprout or germinate. The sprout breaks through the seed coat, and the future stem shoots upward into the light and air, and the root turns downward from them.

153. As soon as the stem rises above the surface it commonly spreads out two seed leaves, which had been already formed in the seed. These leaves, or **Cotyledons,** may be always seen in a bean, pea, or apple seed, which has just come up. But none of the grains or grasses have them. The cotyledons are quite unlike the succeeding leaves of the plant. It is important to remember this, as we often want to know both cultivated plants and weeds as soon as they are up.

154. Plants which have two seed leaves or cotyledons are called **Dycoteledonous** (from two Greek words, *dis* and *cotyledon*, meaning *two-seed-leaved.*) In plants of this kind there appears, between the seed leaves, as soon as the plant is up, a little bud of unopened leaves called

155. The **Plumule.** This soon begins to stretch upwards, bearing on its summit one or two minute leaves nearly

of the usual shape. These enlarge and expand, and from their axil or inner angle, appear one or two other, ordinary leaves, which, with the new joint of the stem, rise and expand in like manner.

156. But all plants do not have two seed leaves. A kernel of maize or of wheat has only one cotyledon. This is also true of all the grains and grasses and of some other plants. Such plants are named **Monocotyledonous** Plants, (plants with *one seed leaf.*) A plant of this kind comes up with one single leaf rolled together, as may be seen in the case of Indian corn or common wheat. When this leaf is somewhat expanded, another leaf appears within it, growing from a second joint in the stem. From each successive joint grows one leaf, till the corn-stalk or grass-stem is complete.

157. The stem of a tree has external and internal organs. The external are the trunk, the boughs, limbs or arms, the branches, the branchlets, the spray, and the shoots or twigs.

The *trunk* is the main body of a tree. It begins at the collar, and, after rising to a greater or lesser height, divides into branches or ramifications. All the divisions, large and small, are called branches or boughs. The largest are called also limbs or arms. A division of a branch is called a branchlet; and all the smallest divisions together are called the *spray.* Shoots or *twigs* are, properly, those of not over one year's growth.

158. A shoot begins in the spring to grow from a bud at the end of a branch called a terminal bud, or from an axillary bud, or one in the axil of last year's leaf, that is, the angle above the leaf, between it and the stem.

159. The internal organs are the inner bark, in several layers, the alburnum or sap-wood, the heart-wood, the pith, &c.

160. The usual course with plants is to grow up, bear leaves and flowers and finally fruits, and then, if they are plants of a single year, to die; if plants of two years, to die down to the ground; if plants of many years, with woody stems, to shed their fruit and leaves, after having formed buds, out of which shall grow the leaves, flowers and fruit of the next year. Those which die at the end of one season, like wheat and Indian corn, are called *annual* plants. Those that live only two years, like beets, carrots and most other garden vegetables, are *biennial;* those that live many years, like shrubs and trees, are *perennial* plants.

161. It sometimes happens with different kinds of cultivated grains, and some other plants, that the plant dies and falls before the seed is quite ripe. Foreseeing this, the husbandman reaps or mows grains and grasses before the seed is ripe, dries them in the sun and air, and leaves them, in sheaves or stacks, completely to ripen their seeds. He thus saves many grains and seeds which would otherwise fall upon the ground and be lost.

162. As the kinds of plants are almost innumerable, they must be arranged in divisions, classes and families, so that they may be studied and recognized. How are they classed? All plants with flowers belong to one or the other of the two great classes just now mentioned, Monocotyledonous and Dicotyledonous.

163. Botanists, since the time of Linneus, until recently, have followed him in dividing plants into classes and orders, made with reference to the number and situation of the stamens and pistils. This is called the *Artificial system* of Linneus.

164. Plants are now best divided into *natural families*, according to the resemblance or analogy of all their organs.

All those which seem to be made upon the same plan, with similar stems, leaves, flowers and fruit, are said to belong to the same **Natural Family.** Thus all the oaks, chestnuts, beeches, and hazel nuts, belong to the *Oak Family*, because, while they resemble each other in general appearance, in the structure of their flowers and fruit they are still more strikingly alike.

165. Plants are still farther divided into *genera* and *species*. A *genus* is a subdivision of a *family*, and a *species*, a subdivision of a *genus*. The oak family, for example, is divided into the *genera*, oak, beech, chestnut, hornbeam, hop-hornbeam and hazel. The genus oak is subdivided into white oak, red, black, post, over-cup, live, willow, and many other species. Speaking of a black oak, we should say; it belongs to the Class Dicotyledonous Plants, to the Oak Family, to the genus Oak or Quercus, and to the species Black Oak, or Quercus Tinctoria.

166. An example will show of what practical use these divisions and subdivisions are. I find a grass which I suspect to be Common Hair Grass; I wish to know certainly; and turn to a volume (Gray's Manual of Botany) which contains a description of every plant in New England. The first part of the volume is occupied with dicotyledonous plants. I find the description of monocotyledonous plants, to which I know grass belongs, beginning on the 426th page. Not desiring to read the whole of 158 pages, I look for the Grass Family, and find it to be the 134th family, and on the 535th page. This family, I find, contains 65 genera. After some examination of a table, I find that the 47th genus of grasses is Hair Grass, (*Aira.*) Carefully reading the description of the genus, in six lines, and of the first species, (*Aira flexuosa,*) in four, I find that the plant I have found belongs to it, and

is, really, Common Hair Grass. Thus, if I understand the language of botany, I can find, in a few minutes, by means of these divisions and subdivisions, what I should otherwise have to read a volume through to find.

Besides, when I have studied one plant of a family and know all about it, I find I thereby already know a good deal about every other plant of the same family.

167. It will be useful to the farmer to know the names of some of the natural families to which the more important cultivated plants belong.

All the kinds of pea, bean, tare, vetch, clover, lucerne, &c., with flowers more or less resembling a butterfly, (*papilionaceous*,) belong to the **Pulse Family,** pod-bearing or leguminous vegetables. The seeds of all these are nutritious to man, and, with their leaves and stems, are of great value to the domestic animals.

168. The cabbage, turnip, radish, mustard, pepper-grass, water-cress, charlock, &c., belong to the **Cress or** ***Cruciferous***, (cross-bearing,) **Family,** so called because their flower-leaves form a cross. To the same belong many plants cultivated for the beauty of their flowers, stock, wall-flower, rocket, sweet alyssum, candy tuft, &c.

169. Flax belongs to the **Flax Family,** valuable in the arts.

170. The roses, peaches, apricots, plums, cherries, hawthorns, apples, pears, quinces, as well as brambles, strawberries and many other plants, with flowers which are like a little rose, belong to the **Rose Family.** The fruits of all these plants are wholesome; many of them, very delicious.

171. Cucumbers, squashes, pumpkins, and melons belong to the **Gourd Family,** with some exceptions, an innocent and valuable family.

172. Currants and gooseberries, both cultivated and wild, belong to the **Currant Family,** whose fruits are healthy and often medicinal.

173. The carrot, parsnip, caraway, celery, parsley, coriander and others belong to the **Parsley Family,** *Umbelliferæ*, (umbel or umbrella-bearing,) so valuable for their roots or their seeds.

174. The sunflower, Jerusalem artichoke, succory, salsify, dandelion, lettuce, daisy, mayweed, chamomile, aster, golden-rod, thistle, everlasting, and many others, belong to the **Sunflower** or **Composite Family.**

175. Sage, mint, sweet basil, lavender, pennyroyal, balm, catnip, hyssop, summer savory, marjoram, thyme, motherwort, horse-mint, spear-mint, self-heal, and many other herbs, belong to the **Sage** or **Mint Family,** friendly, soothing, and pleasant to man.

176. The sweet potato, morning glory, convolvulus, and others, to the **Convolvulus Family,** a suspected race, whose roots are, notwithstanding, sometimes of great value.

177. The tomato, potato, capsicum, petunia, stramonium, henbane, tobacco, &c., belong to a very poisonous family, called the **Night-shade Family.** The root even of the useful potato retains some of the characteristic poison. This poison may always be boiled away. A potato should therefore be so cooked as to be mealy. The waxy appearance shows that some of the poison is still present.

178. The lilac, privet, fringe-tree and ash belong to the **Olive Family.**

179. All the whortleberries, blueberries, cranberries, the checkerberry, May flower, Kalmias or American laurels, azaleas, and many others, belong to the **Heath Family.** Of these many are wholesome, some doubtful, some poisonous.

180. The beet, pigweed, or goosefoot, orache, spinach, &c., to the **Goosefoot Family**, a useful but sometimes troublesome tribe.

181. Buckwheat, rhubarb, sorrel, dock, and knotweed, belong to the **Buckwheat Family**, *Polygonaceæ*, some of which are pleasant as food or as a salad, but some are acrid.

182. The black walnut, butternut, English walnut, and the hickories, belong to the **Walnut Family**, which furnishes us with wholesome and delicious nuts, and wood of great value.

183. The birches and alders belong to the **Birch Family**;

184. The willows and poplars to the **Willow Family**.

185. The pines, the larch, the fir, cypress, arbor vitæ, juniper, yew, white cedar, red cedar, spruce and hemlock belong to the **Pine Family**, of great value to builders.

All the above and many other families belong to the **Dicotyledons**.

186. The following belong to the class of **Monocotyledons**. The lilies, asparagus, hyacinth, crown-imperial, onion, garlic, and many others, belong to the **Lily Family**.

187. Narcissus, amaryllis, tuberose, snowdrop, &c., to the **Amaryllis Family**, valued for its beauty, but also furnishing food.

188. Iris, crocus, cornflag, tiger-flower and blue-eyed grass, to the **Iris Family**. This and the next family minister to our love of beauty.

189. Lady's slipper and the orchises belong to the **Orchis Family**;

190. The rushes to the **Rush Family**;

191. The sedges to the **Sedge Family**, good for the basket maker and the thatcher.

192. All the grasses, all kinds of grain of which meal or flour is made, called the *cereal* grains, such as wheat, barley, rye, oats, rice, maize or Indian corn, and also the sugar-cane, broom-corn and millet, belong to the **Grass Family,** the most friendly of all to the family of man.

193. The **Mosses** are low plants with many leaves and a peculiar fruit, like bird-wheat.

194. **Lichens** are the thin crust-like plants which we see covering the surface of rocks, trunks of old trees, &c.*

195. The difference between a tree, a shrub, and an undershrub, is not precisely marked. A tree is taller than a shrub. Most of the oaks are trees; but two of those growing in New England are shrubs. Most shrubs throw out branches very near the ground, but some, the sweet fern, for example, usually do not. Undershrubs are very low shrubs, like the low blueberries, cranberries, and pigeon plums, checkerberry, and May flower.

196. For the cultivation or planting of perennial plants, the soil must be stirred as deeply as can well be done. Annual plants do not throw down their roots so far into the earth, and therefore do not absolutely require so deep cultivation. But most of them repay the expense and trouble of deep ploughing; and annual plants, having but a short time to grow, must be supplied with a great abundance of suitable food.

197. Some plants are cultivated on account of the value of their seeds, roots or fruits, as food for man. These are called *alimentary.* Others are cultivated as food for

* For full and exact information upon the whole subject of plants, their growth, structure, names and properties, study a delightful little book by Prof. Asa Gray, called HOW PLANTS GROW. For still fuller information, study Gray's LESSONS ON BOTANY. For the fullest and most philosophical information to be found in any one volume in our or any other language, study Gray's STRUCTURAL and SYSTEMATIC BOTANY.

6

other animals, and may be called *forage* plants; others, to yield materials for use in the *arts*, to furnish oil, sugar, dyes, &c.

198. Of the origin of some of the cultivated plants very little is known. Wheat is not now found in a wild state; and the same is true of most of the cereal plants. Indian corn is known to be a native of America, and is thought to have been first carried hence to the Eastern continent.

199. Those cultivated plants which are to be found in a wild state, have been greatly improved by cultivation, especially by giving them a full supply of all the food they need. The wild carrot has a hard, slender root, containing very little nourishment. The cabbage found wild on the coast of France is a small, sharp-tasted plant, without any of the excellent qualities possessed by the different sorts of the cultivated cabbage.

The potato, which is found growing spontaneously in the mountains of Peru, and in other parts of America, has there green, bitter, unwholesome tubers, no larger than a chestnut.

The most striking improvements have been made by the arts of cultivation, by richness of soil and abundance of food, in the fruit of the apple tree. The original tree from which all the others have been derived, is by some persons supposed to be the crab-apple tree, whose fruit is very small and very sour.

200. The size, sweetness and other excellent qualities of most cultivated plants are thus owing in a great degree to the art and care of the gardener and the husbandman, and would lose those qualities if they were long suffered to remain neglected,—left to themselves.

The same seems to be true of all the animals which are subject to man. Their most valuable qualities have been, in a great degree, produced by the intelligent care of men. The same is true of man himself. Children suffered to remain uncared for and neglected,—left to themselves,—are likely to grow up in a condition little better than that of savages.

CHAPTER VII.

ELEMENTS OF PLANTS.

201. The chemists have found, by careful examination, with the help of the microscope, that plant-cells are never formed except in a fluid containing oxygen, carbon, hydrogen and nitrogen. These then are the elements of which all parts of all plants are composed.

Of these, oxygen and carbon are obtained from carbonic acid, and hydrogen and nitrogen from ammonia; and both carbonic acid and ammonia are always found in the atmosphere, and are taken in by the leaves, or are dissolved by the rain falling through the air, and carried into the earth, where they are absorbed by the soil, and hence taken up by the roots.

It may also be that the oxygen and hydrogen are furnished by water, and nitrogen as well as oxygen by the nitric acid sometimes found in the air, and dissolved and brought down by rain.

202. The simplest plant, consisting of only a single cell, must have the power of decomposing carbonic acid, ammonia, nitric acid, and perhaps water.

203. That which causes water, and, with it, these three gases, to enter the plant-cell, is called the **Osmotic Power.**

An experiment which any body can make, shows its action. Let some sugared water, in a tube closed below with a film of bladder tied across the end, and open above, be suspended in a vessel of pure water. The liquid in the tube is soon seen to increase by the passage of the pure water upwards through the film. At the same time, some of the sugared water passes through the film downwards into the vessel.. The tube will soon be full and flow over into the vessel, and the double action will continue till the liquids inside the tube and outside are of the same sweetness and density.

The passage of the fluid from without inwards is called *endosmose;* that from within outwards, *exosmose.*

Two gases, of different density, separated by a film, will, in the same manner, pass through it and mingle.

It is by this power that the various substances that enter a plant not only pass into the cells but also from cell to cell, through all parts of a plant. It is by this, perhaps, that the gases find entrance through the leaves and the tender bark of recent twigs. It is by the same power that fluids are thought to pass from cell to cell, through membrane after membrane, in the bodies of animals.

204. Every part of a plant, even the solid wood, contains **Water,** not always in a fluid state, but in such a state that the chemist can separate water, or the elements of water, even from the dryest wood or bark. Water must therefore be supplied to growing plants in abundance, according to the nature of the plant and the season of the year. Without it, in some form, no plant can grow.

205. **Carbonic Acid** is the most indispensable and abundant article in the food of all plants. It enters the plant dissolved in water, and either remains in that state, or the vital action of the plant, in the light of the sun, decomposes the acid, and throws back most of the oxygen into the atmosphere; but retains a portion which performs important offices; and also retains the carbon. This forms the solid parts of every plant. The walls of the cells, the wood, the frame-work of the leaves and of every other part, are made of carbon, together with oxygen and hydrogen in the proportions in which they form water.

206. Hardly less important to the nourishment of plants is **Ammonia.** This is a gas of a very pungent odor and burning taste, which, when absorbed by water, forms what is commonly called *spirits of hartshorn.* It has a great attraction for carbonic acid, with which it combines and forms *carbonate of ammonia*, popularly called *smelling salts.*

Ammonia is composed of hydrogen and nitrogen; and as both these substances are always found in living plant-cells, and must be essential to the life and growth of these cells, not less essential is ammonia, or some other source of nitrogen, such as nitric acid.

207. Carbonic acid, ammonia, nitric acid and water, obtained thus from the atmosphere, are the atmospheric food of plants, and the four simple elements which they contain, are the only ones always found in every plant, and therefore considered absolutely essential.

208. From the fact that these essential elements are derived from the atmosphere may be understood the possibility of the growth of air-plants, which flourish

without any immediate connection with the earth, and drink in all their food from the air.

209. The charcoal in plants is never found perfectly pure. Diamond is pure carbon. In plants it is always combined with something else. By charring, that is, exposing wood or other vegetable substance to great heat, out of the reach of the open air, all the atmospheric portions are consumed, or, to speak more properly, turned into vapor and gases, and driven off, and a perfect skeleton of charcoal, showing all the minutest parts of the structure of the plant, is left.

210. In peat, which is the woody substance often found under the surface in swamps, and also in anthracite and bituminous coal, which are the remains of the vegetation of former ages, every thing in the structure of the plants, of which these substances are formed, is often so completely retained, that from them the family, and even the genus and species of the plant may be ascertained.

211. By the process of charring, every thing except the carbon is not consumed. Indeed **nothing is consumed;** but those portions capable of assuming a gaseous form are driven off. By carefully burning, in air, the charcoal left, the carbon combines with the oxygen of the air and flies off in the state of invisible carbonic acid, a portion of water which has still adhered to the charcoal is turned into vapor, and a greater or less amount of *ashes* is left.

212. All those elements which thus assume a gaseous form and fly off into the atmosphere, as smoke, vapor or gas, in these two kinds of burning or *combustion*, are often called for that reason, the *combustible*, or, more properly, the atmospheric *elements*. They are oxygen, hydrogen, carbon, and nitrogen, and their compounds,

water in the state of vapor, ammonia, carbonic acid and some others.

Those that are left in the **Ashes** are the *incombustible* elements, or the mineral elements. In the ashes of every plant is found a very considerable number of mineral constituents. But the ashes of plants of particular families are often remarkable for the amount of particular elements contained in them.

213. The ashes of radishes, mustard, and other plants of the *Cruciferous Family*, particularly of the seeds, contain **Sulphur,** or brimstone, in the state of *sulphuric acid*, combined usually with some other substance.

214. In the ashes of pod-bearing or leguminous plants, such as peas and beans, and other plants of the *Pulse Family*, particularly clover, sulphuric acid in composition with lime, or **Sulphate of Lime,** is found.

Lime is a compound of a metal called calcium, with oxygen; so that *sulphate of lime* is made up of sulphur, oxygen and calcium. It is commonly called gypsum, or plaster of Paris.

215. In the ashes of kernels of wheat or other grain, as well as of many other kinds of seed, is found a large quantity of a salt called *phosphate* of lime. This is a compound of lime and *phosphoric acid*, which is itself composed of oxygen and a very curious substance called

Phosphorus. This is a soft, translucent, poisonous solid, looking like wax, turning yellowish when exposed to light, of a peculiar smell, and called phosphorus, (light bearer,) from shining in the dark. It has so violent a tendency to combine with the oxygen of the air, and burn, that it must be kept under water. A very little of it mixed with other substances and applied to the end of a bit of wood, gives that readiness to take fire which belongs to phosphorus matches, commonly called lucifer or friction matches, which a little rubbing produces heat enough to set on fire.

Phosphate of lime is found not only in the *seeds* of very many plants, especially those of which bread is made, but in all plants, and in the bones of men and other animals, whence it is called *bone-earth.*

216. The ashes of all kinds of straw and grass, of the bamboo cane, and of the scouring rush, consist, in a very large degree, of silex or *silica;* and all these plants owe the stiffness and hardness of their stems to the silica contained in them.

Silica is oxygen combined with a metal-like substance called silicon. When perfectly pure, *silica* is a white, gritty powder, without taste or smell. It is the substance of which quartz, rock-crystal and flint are composed. Though wholly unlike, in appearance, to the other acids, it is yet an *acid*, and combines with the oxides of many of the metals to form *silicates*, and, in these forms, constitutes a very large portion of all rocks and soils.

217. In the ashes of trees and other woody plants, as well as in most other ashes, *potash* is found. If wood ashes be leached, that is, if hot water be poured upon them, it will, in a short time, dissolve the potash in the ashes. The dark-colored, strong lye, thus obtained, boiled with oil or fat, forms common soft *soap.*

Lye, boiled away, in a pot, without fat, leaves a dirty-looking substance called *potash.* This, when somewhat purified, is called pearlash.

218. This common **Potash** is the carbonate of potassa, a compound of carbonic acid and potassa, which is, itself, a compound of oxygen with a metal called potassium. This metal has the lustre of silver, but is soft, and so light

as to float on water. So great is the attraction between potassium and oxygen, that it decomposes the water on which it floats, unites with a portion of its oxygen, exhibiting the singular appearance of a little fire on the water, and forms potassa.

219. In the ashes of kelp and of other plants growing in the sea, and of some of those growing near the sea, instead of potash, **Soda** is found, in the state of carbonate of **Sodium,** a light metal somewhat similar to potassium, and having nearly the same violent affinity for oxygen, so as to take fire when placed on hot water.

220. **Alkali.** The ashes of sea plants have long been of value in commerce, from being used in the manufacture of hard soap, and also of glass. These soda ashes are called, in Spain, alkali, (Arabic *al*, the, *kali*, ashes,) which name has thus been given to *soda*, and thence to *potash* and *ammonia*, all which are called *alkalies;* and all three have very similar properties. They have a bitter, acrid and burning taste, and the power of changing vegetable blue colors to green, and pink to blue.

221. They have also the remarkable property of uniting with the acids, and thereby losing all their own peculiar properties, and destroying those of the acids. Sulphuric acid, for example, has the extreme sourness and corrosive power with the other properties of the acids. Pure soda has the alkaline properties just mentioned. But when sulphuric acid is poured upon soda, it forms a new substance, sulphate of soda, or Glauber's salts, which is called a neutral salt; a salt, because it looks very much like common table salt, and neutral, because it has *neither* the properties of an acid nor those of an alkali.

It is in the state of neutral salts that most of the mineral substances enter into the composition of plants.

222. The ashes of asparagus, and of other plants which grow naturally near the sea, contain a large portion of common salt, in very minute, regular, cubical particles, called crystals. Now salt is composed of a gas called *chlorine*, and of the metal *sodium*, and this salt,—common table salt,—is called by the chemists **Chloride of Sodium.** And it is very remarkable that this pleasant and wholesome article in our food should be composed of a substance so ready to take fire as sodium and another like

223. **Chlorine.** This is a suffocating and poisonous gas, of a greenish color, whence its name, (chloros, Greek for green,) which has a great attraction for foul air and for coloring substances, and is therefore employed for disinfecting, or drawing off foul air, and for bleaching, or making things white.

224. Oxides of two other metals, **Magnesium** and **Iron,** are also found in the ashes of all plants, but commonly united with some one of the acids.

225. The oxide of magnesium is called **Magnesia.** It is a white, bitterish substance, resembling flour in appearance, often used in medicine.

226. Plants growing in the sea, called sea-weeds, such as kelp, oar-weed, rock-weed, &c., and those growing on the sea-shore, contain, in their ashes, salts of two substances, called *iodine* and *bromine*.

227. **Iodine** is a solid which looks like black lead. When heated, it throws up a violet colored vapor, whence its name, from a Greek word, (i-o-des,) meaning violet-colored. If a polished silver plate be held over this vapor, it becomes first of a yellowish color, then violet, then deep blue, from the combination of the iodine with the silver. This compound is powerfully acted upon by light, and hence its use in the processes of the daguerreotype.

228. Iodine occurs in plants as *iodides*, or compounds of iodine with some metal, as, for example, the iodide of potassium. Bromine is found in a similar state, that is, as *bromides*.

229. **Bromine** is a heavy, brownish liquid, of a suffocating odor. When scarcely perceived, this odor is not unpleasant, and this, with the odors of iodine and of chlorine, forms probably the pleasant smells we perceive on a sea-beach.

230. These are the principal and the most important mineral substances found in vegetables.

But a metal called **Aluminum**, which is the basis of clay, and also the metals **Manganese** and **Copper** are found, very rarely, in the ashes of some plants.

231. Are all the substances necessary to the growth of a plant, of equal value? All are essential. If any one of the whole number be absent, the plant will not thrive; but all are not needed in the same quantities.

232. The **Acids** most important in the structure of plants are *carbonic acid*, *sulphuric acid* and *phosphoric acid*, either by themselves, or united with substances with which they form salts, such as *carbonates*, *sulphates* and *phosphates*. These are found in all plants. *Silicic* acid combined with the alkalies and with the earths is also essential to very many plants.

233. But these are not the only acids found in plants. By a peculiar action of the vital power of particular plants, the elements of carbonic acid and water form a variety of acids differing from carbonic acid and from each other.

The acid which gives to apples their characteristic taste, is called *malic acid* (Lat. *malum*, an apple.) The acid of oranges and lemons is *citric acid*, (Lat. *citrus*, an orange;)

that of wood sorrel (oxalis,) *oxalic;* that of grape vines and grapes, *tartaric acid.*

234. All these unite with the oxides of the metals that have been spoken of, and one or more of the salts formed by the union are found in the cells or at least in the ashes of nearly all plants. The salts of potassa, for example, are always found in the ashes of potatoes, turnips, the grape vine and many others; and none of these plants can flourish in a soil, however rich in other respects, which contains no potash. Hence potatoes, turnips, beets, and Indian corn, are sometimes called **Potash Plants.**

235. In like manner oats, wheat, barley and rye are called **Silica Plants,** because the ashes of the straw of these plants are more than half made up of silica. And because tobacco, pea-straw, clover, and potato-tops, leave ashes of which more than one half is lime, these plants are called **Lime Plants.**

236. **Phosphates of Lime and Magnesia,** in small quantities, are found in the ashes of all common plants; but they form from one half to three fourths of the ashes of wheat, and a very large portion of the ashes of other grains.

237. What then are the most essential elements in the growth of plants? All plants, without exception, require for their subsistence and nutrition, the atmospheric elements, oxygen, nitrogen, hydrogen, and carbon, and the earthy elements, phosphorus, sulphur, potash, lime, magnesia, and iron. Plants of certain families require silica. Others require common salt, soda, iodides and bromides.

238. Besides these, three metals, aluminum, manganese, and copper, are found very rarely, as oxides, or as salts, in the ashes of a few plants; and, still more rarely, **Fluorine,** a powerful gas, remarkable for its power to

corrode glass, is detected in the ashes of some plants. It occurs in combination, as *fluoride of calcium*, or fluor spar, in which form it is also found in the teeth and bones of animals.

All these earthy substances are called the mineral food of plants.

CHAPTER VIII.

ORGANIC COMPOUNDS IN PLANTS.

239. Of the simple, elementary substances spoken of in the last chapter, and their direct compounds, although they are all found in plants, none ever appear in particles large enough to be seen by the naked eye. Of them, however, are formed the substance and the nutritious and other useful products of the plants, called the *organic compounds.*

240. They are so called, because they are compounds formed by the action of the vital power of the organized being, a plant.

241. Among the most important are, first, those formed of carbon, oxygen and hydrogen only, viz., Cellulose, Vegetable Jelly, Starch, Gum, Sugar, and Oil.

242. **Cellulose,** also called woody fibre, is the cell-membrane, or thin covering of the cells. When first formed, it is tender, flexible and elastic, clear and transparent. It is expanded by moisture and contracted by drying. It is permeable to all fluids, which enter on one side and pass out on the other. It is called woody fibre, because it

forms the substance of all wood, giving it strength, hardness and elasticity.

243. **Vegetable Jelly** is so called, because, while moist, it looks and feels like common jelly. When dry, it becomes horny or cartilaginous. Quince jelly and apple jelly are forms of it, but mixed with the acids and other compounds which give them their peculiar taste.

244. Every-body is familiar with the appearance of **Starch.** When dry, it is somewhat hard, and crumbles between the fingers. When moist, it is somewhat like jelly. It is completely soluble in warm water, and, when perfectly pure, is clear and transparent. As it dries, it is at first a trembling jelly, but at last becomes brittle as glass.

Starch is found, already formed, in almost every plant that has been examined, particularly in the grains of all the cerealia, in beans and pease, and almost all seeds, in potatoes and all other esculent roots, and in the pith of many plants, as in the sago palm. In arrow-root it seems to be purest. Starch, variously compounded, but never absolutely pure, constitutes the most important, and often the only food of two thirds of all mankind. It occurs in small quantities in the bark and newly formed wood of many trees, in winter, whence the inhabitants of the Polar regions are able to use the bark of trees, when baked, as bread. It is extracted, for use in the arts, from potatoes, wheat, and some other substances.

245. **Gum** is the substance which we often find hardened in roundish masses on the bark of cherry and peach trees. It is in all plants; in plants belonging to some families, it is found very abundantly. Gum Arabic is a well-known form of it. When pure, it is clear and transparent; when dry, very brittle. It easily dissolves in water and in weak acids, but not in alcohol. It is very nourishing,

and is sometimes used as food. By the botanists, one form of it is called dextrine.

246. Loaf sugar is **Sugar** in a crystalline state. Attentively examined, it is found to be made up of little bright crystals, which reflect the light and give the brilliant white appearance of loaf sugar. Dissolved in water and allowed to evaporate and harden, it becomes sugar candy. Brown or Muscovado sugar is unrefined, and contains other substances which give it its peculiar taste. Sugar is nutritious, and is used, all over the world, as a sweetener. It is found in every plant; but in the greatest abundance in sugar cane, Indian corn-stalks, sorgho, beet root and carrot, and in sweet fruits, as the pear, and apple, and the melon.

247. **Vegetable Oils.** The peculiarity of these substances is their leaving upon paper or linen a translucent spot, and their refusal to mix with water. There is perhaps no plant and no part of a plant which does not contain oil. From some plants, as from a species of palm in Africa, it is extracted in vast quantities. From many seeds it may be pressed, as particularly from the seeds of flax, when it is called linseed oil, and from those of the turnip, the poppy and the sunflower. A plant called *colza*, which botanists suppose to be the cabbage in its natural condition, is extensively cultivated in France for the purpose of yielding oil.

248. **Wax** is a kind of solid oil which often appears on the surface of the stem, leaves or fruits of plants, and in a very remarkable manner upon the fruit of the candleberry myrtle. In those parts of plants which have a hoary appearance, as is the case with many kinds of plum, the delicate bluish bloom consists of a thin layer of very small wax granules. *Bees-wax* is collected, perhaps formed, by bees. Some chemists think it is formed from sugar.

249. All these organic compounds are very nearly related, and often change from one into another. Cellulose may turn into starch, gum or sugar. So may vegetable jelly. Gum or dextrine may be converted into sugar. These substances appear to go successively through all these forms, from sugar, the most soluble, to cellulose, the most insoluble. All these substances, 241, taken into the animal system, are supposed to aid in the process of breathing, and keeping up the warmth of the body.

250. There is another class of substances found in plants, of which the cell-walls are not formed, and which yet are essential to the simplest processes of vegetation. They are composed of the elements of water, of carbon, and also of nitrogen, to which are sometimes added phosphorus and sulphur. From the nitrogen contained in them, they are often called **Nitrogenous Compounds.** In their simplest form they are composed of the four atmospheric elements only, and are found in a fluid, semi-fluid, or solid state, within the cells; and without their presence in a liquid state no new cells can be formed. From their great variety of appearance, and the readiness with which they change, these substances have been called **Protein,** from the name of an imaginary being, Proteus, who was fabled to assume every variety of form, to conceal himself.

251. Protein, in combination with sulphur, forms *casein,* with still more sulphur and a little phosphorus, *albumen,* and with more both of sulphur and phosphorus, *Gluten* or *Vegetable Fibrine.* These substances are of great importance, and of the highest interest, from the fact that though essential to the bodies of animals, constituting the muscles and giving them strength, they are not, according to some chemists,* formed in the animal

* Liebig, and others.

economy, but must be taken into the system already formed.

252. **Casein** is an essential ingredient in milk and in cheese, whence its name (caseus, Latin, cheese.)

253. **Albumen** is nearly identical in composition with the white of an egg (of which albumen is the Latin name) and is found in many parts of the human body and the bodies of other animals. It is always found dissolved in the sap and juices of living plants.

254. Wheat contains from 8 to 35 per cent. of **Gluten,** Indian corn 12, beans 10, rye 9 to 13, barley 3 to 6, oats 2 to 5, potatoes 3 to 4, and a little is found in beets, turnips and cabbages.

255. The fact that wheat varies so much in the gluten it contains is one very instructive to the farmer. When fed with the very richest manures, especially those containing animal substances, wheat not only yields more abundantly, but the grain is richer in this most nourishing element. For *Animal* **Fibrine** is the essential portion of the fibrous part or muscle of the flesh of animals.

356. The elements of every thing in the body of a man or any other animal must have come into the system in the water or air, or in the vegetable and animal food which he has consumed. To exist in the body of an animal, they must have been found in the vegetables on which it has been nourished, and, before that, in the soil out of which the vegetables grew, or in the atmosphere by which all have been surrounded.

257. **What** it is which **gives color to the leaves of Plants.** The green color is owing to a substance called **Chlorophyl,** (leaf-green.) This is found in the leaves and in the bark of the newly formed twigs of nearly all flowering plants. It is composed of a white, wax-like substance, and a

peculiar, green, coloring matter. This green coloring matter is formed under the immediate action of light, and its depth of color seems to depend upon the intensity of the light. Hence the innumerable shades of green, from the delicate yellowish green of early spring to the deep greens of midsummer; and hence the striking changes in the color of leaves, after some days of cloudy, warm weather, when succeeded by clear sunshine.

258. The yellow leaves in autumn contain proportionally more wax than the green leaves of summer, and the yellow rinds of ripe fruits more than the green rinds of unripe fruits. The rich, gorgeous colors of the autumnal foliage have been attributed to the action upon chlorophyl of various vegetable acids and alkalies, under the influence of the sun's light. They are not produced by frost.

259. From the roots, wood, bark and leaves of various plants are extracted very many *coloring substances* used in the arts. Certain plants, as, for example, the indigo plant and woad, are cultivated extensively, in some countries, for this very purpose.

260. **Tannin.** This is the substance with which tanners convert the hides of animals into leather. It is found in the bark of several kinds of oak, and also of hemlock, spruce and some other trees of the Pine Family, and in the leaves of tea and of some plants of the Heath Family. It is of a sourish, astringent taste, and has this remarkable property of converting the animal gelatine of the skin into leather. Tannin is found only in the older wood and bark, and is supposed to be formed by the commencement of decay in cellulose.

261. How the vital principle in plants, with the agency of the osmotic power and chemical attraction, forms the

various products which have been spoken of, and innumerable others, we can only conjecture.

Some of the imagined operations are strikingly set before us in a picturesque passage which may form a fit conclusion to this chapter.

"The vessels of vegetables have the same wonderful, and seemingly intelligent power of selection, that exists in the vessels of animals. They are thus enabled to select from the compound circulating sap, what each set of vessels requires, to construct the tissue which each has in charge. One set selects materials for the alburnum, another for the bark, another for the leaf and the leaf-bud; another forms the fruit-bud, and ultimately builds up the fruit. One set constructs the woody-fibre, another set the starch, another the gum, another the resin, another the bitter principle, another the sweet and nutritious juices, another the poisonous elements. One set forms from the sap, the coloring matter that blushes or glows in the petals of the flowers, and the coverings of the fruit. Another selects, atom by atom, the lime that enters into the composition of the grain of wheat; another set weaves the covering for this same grain, from the woody fibre. Another set deposits the fatty elements, and arranges them in layers, around the starch and sugar and lime, of which the kernel of corn is built up. Thus every tissue and every product of vegetable life are formed by innumerable vessels, from the descending sap."*

* See a beautiful "Prize Essay" upon Manures, by JOSEPH REYNOLDS, M. D., of Concord, Mass.

CHAPTER IX.

THE SOIL.

262. Of the vast interior of the earth nothing is known with absolute certainty. We are acquainted with the outer portion, the crust, only; and the geologists and the chemists have been studying that very attentively for many years.

By this careful and continuous study, the crust of the earth, together with the waters resting upon it and the atmosphere enveloping it, is found to be made up of sixty-one, perhaps sixty-two or sixty-three, *elements*. Several of these, when pure, are gases; but all are found, usually in combination one with another, in a solid state. Several of them may possibly be hereafter found to be formed of one and the same substance.

263. All these elements, except twelve or thirteen, are metals, more or less like iron, copper, lead, tin, mercury, gold and silver. The greater part of them are *found* only as *ores*, that is, combined usually with oxygen, or with sulphur, carbon, or something else, and often looking like earths, which indeed they are. About thirty-four of them are found in very small quantities, and are seldom seen except by chemists.

264. Only a few, as gold, silver, copper, mercury, and platinum, are found in their native state, in the earth, in the condition of purity. Metallic masses and fragments of stone, called *meteoric* stones, or *aerolites*, are sometimes seen to fall, and are always supposed to have

fallen, from the sky. These are often found upon or near the surface, and consisting of iron and two other pure metals not oxidized, in the form of a brilliant, malleable compound. All the rest, whether found at the surface or deep beneath, are in rocks or the fragments of rocks.

265. The study which searches into the structure of the earth, asks what the rocks are and in what order they lie, and examines the curious remains of plants and of animals that are often found in them, is *geology;* and a person who pursues this study is a *geologist.*

The study which searches into the inner nature of things, to find out what they are, what they are made of, and how they act on each other and on animals and plants, is *chemistry.* A person who pursues this study, with experiments, is a *chemist;* and the process of searching, by experiments, and separating a compound substance into its *elements*, is *chemical analysis.*

266. The **Soil** is that part of the ground which can be tilled, which can be reached and stirred by agricultural tools. It is made up of many different kinds of earth. Of these the three most important are *silicious* earth or sand, *argillaceous* earth or clay, and *calcareous* earth or that made of limestone or carbonate of lime; and, by the mixture of these three, most of the different kinds of soil are formed.

267. The soil which covers the surface of the earth rests upon rocks lying at a greater or less depth beneath, from the crumbling or disintegration of which the soil and loose earth have apparently been formed.

The principal and most important of these rocks are the following: first, Granitic Rocks, including Greenstone Rocks; second, Silicious Rocks; third, Slaty or Argillaceous Rocks; fourth, Pudding-stone Rocks; fifth, Limestone or Calcareous Rocks.*

268. (1.) The **Granitic Rocks** get their name from **Granite,** which is a hard rock composed of three minerals called quartz, felspar and mica. **Sienite** is like granite, but is composed of quartz, felspar, and hornblende; and **Greenstone** is composed of felspar and hornblende, without quartz. **Traprock,** another very hard rock which often forms what seem to be natural walls, sometimes with steps in their ends, is composed of felspar and hornblende, with another mineral called augite. **Gneiss** and **Mica Slate,** which look and are exceedingly like granite, consist chiefly of mica and quartz, with felspar; and **Porphyry** is a very hard rock, made up almost entirely of felspar.

269. Granitic Rocks, including all those mentioned above, are extremely hard, and are thought to be among the oldest rocks. They, or the minerals of which they are made up, are chiefly composed of 1, *silex;* 2, *alumina;* 3, *lime;* 4, *potash;* 5, *magnesia;* and 6, *oxide of iron;* and, by their crumbling, or disintegration, form granitic earths.

270. Far the most abundant of these six is silex or silica, which, as we have already said, is a metal-like substance, *silicon,* chemically united with oxygen. Though it is not sour, it has other properties of an acid, acts as one, and is called *silicic acid;* and the other five sub-

* The teacher should, if possible, be furnished with a small collection of specimens of rocks and of the more important minerals found in them. By means of these his instructions may be made far more interesting and intelligible than they possibly can be without. For perfect illustration of what is taught in this chapter not more than twenty specimens will be required; and, by means of such a collection, the pupils may easily be induced to make collections for themselves, and to become acquainted with the names and qualities of all the rocks in their neighborhood.

stances mentioned above, are usually combined with it as silicates of potash, silicates of alumina, &c.

When found pure, it is called quartz or flint, and in that state is used in the making of glass. It is the most abundant solid constituent of the earth's crust, forming about five-eighths of the substance of the most important rocks. Agate, chalcedony and opal, which are hard and almost precious stones, are nearly pure silica. Though so very hard, it is rendered soluble, and is dissolved by the action of the alcalies and their carbonates.

271. Silica usually occurs as coarse or fine sand, and enters very largely into the composition of the soil of all granitic regions, such as that of the greater part of the New England States. *Pure* silicious sand is seldom found. It is commonly mixed largely with grains of sand formed by the crumbling of the other ingredients of the rocks.

272. (2.) **Silicious Rocks** or sand-stones are composed of small grains of silex agglomerated or stuck together, and of various colors, from white to red, according to the proportion of oxide of iron which they contain. When crumbled into loose sand they make the poorest possible soil.

A soil formed principally of the sands coming from these two sources, is a loose, light, sandy soil, readily penetrated by water, but not retaining it long, and therefore liable to be much affected by drought. It is easily cultivated, but not fertile, especially when its principal ingredient is coarse silicious sand. Its fertility and its readiness to retain moisture and manures depend upon its fineness and upon the due admixture of other ingredients of soil, clay and lime, to be spoken of presently.

273. (3.) **Slaty** or **Argillaceous Rocks** are all more or

less like slate, and, by their crumbling and decomposition, seem to have given rise to *clay* or argillaceous earth.

Clay is silicate of alumina; a chemical compound of silicic acid, alumina, and water. Clay usually contains also silicates of potash, of soda, and of lime. It forms a compact, fatty earth, which is soft to the touch, adheres somewhat closely to the tongue, and exhales a peculiar odor, which is perceived when it or clay-slate is breathed upon.

Pure clay is white; but clay, as ordinarily found, is colored blue, brown or red, by oxides of iron. It absorbs a great deal of water, and parts with it very reluctantly; and it has a strong attraction for ammonia and for the very richest portions of manure.

When completely wet, it becomes a thick paste, almost impenetrable to water and to air, which it prevents from percolating or penetrating farther into the earth. Under the effect of drought, it cracks and becomes excessively hard. From the action of frost, on the contrary, it swells and crumbles into powder, from the water's expanding, as it freezes, and thus breaking up whatever contains it. Hence the usual humidity of clayey lands, the difficulty of ploughing them in a very wet or a very dry season, and the beneficial effects of freezing.

274. There are many kinds of clay, and most of them are of great value in the plastic arts. All the varieties of porcelain, pottery, stone ware, earthen ware, tiles and bricks, are made wholly or chiefly of clay. The celebrated *kaolin*, or pure white porcelain clay of China, is mouldered felspar; and the *petuntze* of the Chinese potter is another kind of felspar containing potash. Clay is also the material commonly used by the statuary, in which to shape the first draught or model of his figures, and often by the

architect for the first solid representation of the ornaments of pillars and other parts of buildings.

275. Most of the slates are more or less aluminous. The metal *aluminum*, which is the basis of clay, very much resembles silver in color, brilliancy and hardness, though far less beautiful. Alum, from which it derives its name, is partly made of it.

The oxide, alumina, is one of the most abundant materials of the crust of the earth, forming not less than one quarter of its substance. Two of the most beautiful of the precious stones, the *sapphire* and the *ruby*, are alumina tinged with a little oxide of iron. They are inferior only to the diamond in hardness and brilliancy. Another very beautiful precious stone, the *topaz*, is also an aluminous mineral. When colorless, it possesses a lustre which has often caused it to be mistaken for the diamond.

276. (4.) **Calcareous Rocks** are composed chiefly of carbonate of *lime*, that is, lime chemically combined with carbonic acid. There also enters into their composition a greater or less proportion of silex or other sand, and of clay and sometimes other mineral substances. In England and some other countries, vast quantities of *chalk*, which is carbonate of lime, are found, and in some places the soil is almost wholly made of it.

A soil consisting chiefly of calcareous earth is a very poor soil. It has more tenacity than sand, but less than clay, absorbs moisture readily, but easily parts with it, and is liable to crack, when dry, like clay, and to parch plants growing in it. Excessive moisture turns it into a thick mud, and if, in this state, it be exposed to extreme cold, it swells and cracks, and is apt to wound the roots of plants and even throw them out of the ground. In its mechanical properties it is a medium between clay and sand.

277. (5.) **Pudding-stone Rocks,** sometimes called graywacke, are made up of materials formed by the mixture of a great variety of other rocks, which seem to have been brought together, in very ancient times, by the action of floods or streams of water. They have their name from their resemblance to plum-pudding, the ingredients being of every variety of lime-stone, clay-slate, and porphyry, greenstone, trap, and every other form of granitic rocks. They are often of a very coarse texture, made up of pieces of stone of every size, sometimes weighing hundreds of pounds, and sometimes of so fine a texture as to resemble slate.

The materials are held together by a natural mortar of lime or of rust of iron, or by mere contact. When completely reduced to dust, these rocks make a rich soil, from its containing all the mineral materials, intimately mixed, which are necessary to the fertility of soil.

278. All these rocks, differing in hardness and in other properties, and forming, perhaps, at first, the surface of the earth, have, in process of time, been crumbled, and then, or before, transported to various distances.

The sand, coarse or fine, formed by the crumbling of the granitic rocks, sand-stones, and pudding-stones, contain the six substances enumerated, 269. The slate rocks form clay; and the chalks and other calcareous rocks, lime. Altogether they furnish all the mineral materials which enter into the structure of plants.

279. How have these rocks been changed into soil? Chiefly by the action of heat, of water, and of cold. The sun's heat warms and expands all the rocks upon which it falls. While they are in this state, the rain, descending,

penetrates their surface and moistens and softens them. Frost turns this moisture into innumerable little wedges of ice, which split the thin outer coat of the rocks into minute fragments. The hardest rocks are thus gradually crumbled into dust.

Besides these agencies, oxygen is constantly acting. So are other gases; and so are carbonic acid and other acids, and lime, and the salts of potash, and other salts. These are dissolving, disintegrating and crumbling the rocks; and water, in streams and torrents, is constantly rubbing off and dashing together the fragments.

All these causes are still and constantly acting, not only upon the surface of the great rocks, but upon the surface of the particles of the soil in the cultivated or uncultivated fields. The ceaseless action of all these and of other forces is called *weathering*.

280. The important question with the farmer is, Which is the best soil? Neither of the three kinds of earth spoken of forms by itself a good soil. Indeed, each, by itself, forms a soil absolutely barren. The best natural soil is one formed by the due mixture of all the three, the bad qualities of each being corrected by the good qualities of the others.

The chemical analysis of a vast number of soils shows that the most fertile are those into which these three important classes of elements enter abundantly, but not in equal quantities; and that the fertility diminishes just in proportion as any one of the three comes near to be exhausted.

281. All the innumerable soils have essentially the same elements. Clay, lime and sand are the basis of all. But soils vary as one or another of these prevails, or as one or another is wanting.

A soil formed by a mixture of clay and sand, in nearly equal proportions, is called a *clayey sand* or a *sandy clay*, according as the one or the other predominates. If much more than one-half is clay, we call it a loamy clay. So we call a soil a *calcareous clay*, or a *clayey calcareous soil*, as the clay or the lime is the more abundant.

282. It must, however, always be understood, that **all these combined, even in the most favorable proportions, are not sufficient to form a good soil.** There must be superadded a certain amount of *humus*, *mould* or *geine*. This seems to be at the same time the reservoir, and often, perhaps, the source, of those saline matters and of a large portion of the nitrogenous and carbonaceous substances which are essential to the growth of plants.

Humus, or **Geine,** for both words mean the same thing, is a dark-colored earthy matter, fatty to the feeling, formed from the remains of vegetable substances, and sometimes also animal, in different stages of decomposition. It readily attracts and absorbs water and retains it, not only rain water but the vapor of the air. It is the perfection of vegetable earth. Land is considered good arable land, which contains three or four per cent. of it. Soil containing as much as eight per cent. of it, is good garden mould, and with ten per cent. it becomes very rich.

283. It can be very readily ascertained whether there is any humus present, by burning a quantity of the soil upon a red-hot fire-shovel. As the humus calcines and turns into charcoal, it exhales an odor either like that of burnt horn or feathers, or like that of burning straw. If the smell be strong of burnt feathers, it indicates a soil rich in the products of decayed animal substance. If the

only perceptible smell be that of burnt straw, it indicates humus formed from decayed vegetable substances.

284. Humus is always favorable to vegetation, except when it has been produced by the decay of plants under water, or has been very long lying under water. This is often the case with peat, bog earth or marsh mud. These are almost entirely humus; but when they have been long beneath the surface of water, they are considered cold, and possess acid properties, which render them unfavorable to the nourishment of plants, until corrected by long exposure to the influences of the atmosphere, and to the alternation of the sun's light and of frost.

285. Humus not only acts as a reservoir of carbonic acid, holding it ready to be given to the roots of plants, but, as it consists mostly of carbon and water, and has an attraction for oxygen, it is constantly receiving oxygen from the air. By the progressive decay thus produced, the vegetable and animal remains are constantly turning into carbonic acid and ammonia, and the ammonia into ammoniacal salts, thus rendering the soil rich in these precious elements of vegetable food. In soil abundantly supplied with humus or other rich manure, the air is sometimes found to contain four hundred times as much carbonic acid as an equal quantity of the air in the atmosphere.

286. The carbonic acid formed in vegetable soil by the oxygen, not only serves directly as food for plants, but it decomposes the silicates and thus sets the potash and other salts free to be dissolved by water and taken up by the roots. Another portion of the *oxygen* absorbed, *combines with* the *hydrogen* of the humus, *and produces water*. This is a very valuable property, especially in dry seasons, and is one reason why soils abundantly supplied with humus suffer so little from drought.

287. Another most important property, and essential to the fertility of soil, is the power of absorbing moisture from the atmosphere. During the night, soils which possess this property in a sufficient degree are enabled to condense a large quantity of water, and thus make up, in a very considerable measure, for the enormous quantity lost by evaporation during the day.

These powers of absorbing oxygen, of absorbing and retaining moisture, and of forming water, are given to a sandy soil by humus, and also by clay, but far more effectually by the two mixed together.

288. **The richest** natural **soils** are those which **contain** all these ingredients, **sand, clay, lime and humus** in due proportions. Such are the alluvial soils found on the low banks of the Connecticut and many other rivers. These streams, in their course from their sources in the hills, wash against and wear away a great variety of rocks, dissolve and carry along with them portions that have been made soluble by the processes of weathering, and take up quantities of leaf and other vegetable mould, and bring them all away in their current. When, in the winter and spring, they overflow their banks, they deposit all these mingled materials upon the intervale or bottom lands,—the low grounds lying between the river and the hills.

In the lower part of a river's course, these various materials are deposited in the state of the finest sand or clayey mud; soils so formed are found to possess an almost inexhaustible fertility. They unite all the materials necessary for the growth of plants, clay, sand, lime and humus, in circumstances the most favorable, all perfectly mixed, and all reduced to the state of the finest powder.

289. Next in value to these soils, for permanent cultivation, are the light sandy soils formed by the crumbling

of the granitic rocks. They contain, in inexhaustible abundance, all the mineral elements necessary to the growth of a plant, potash, soda, lime, magnesia, iron and manganese, in the condition of silicates.

290. The following table will show this to be true. Remember that granite, gneiss and mica slate, are composed of mica, quartz, and felspar; syenite, of quartz, felspar and hornblende; trap-rock, of augite, felspar and hornblende; greenstone, of felspar and hornblende; and porphyry, almost entirely of felspar.

In one hundred parts, there are, in these different minerals, about these proportions. For great exactness, see Dana's Manual.

	Silica.	Alumina.	Potash.	Magnesia.	Iron.	Lime.
In Quartz, . . .	100.	–	–	–	–	–
Felspar, . . .	67.	19.	14.	–	–	–
Mica, . . .	46.	14.	10.	10.	20.	–
Hornblende, . .	59.	–	–	20.	7.	14.
Augite, . . .	53.	–	–	8.	17.	22.

291. Quartz is silica nearly pure. Felspar is a silicate of alumina and potash. Mica is a silicate of alumina and potash, and of magnesia and iron. Hornblende is made of silicates of magnesia and lime, with iron; and augite, of silicates of lime and magnesia, with a larger proportion of iron. In some kinds of felspar soda takes the place of potash.

292. How is a light, sandy soil, possessing the mineral elements of fertility, to be managed, that it may become fertile? The first thing to be done is to render it capable of absorbing moisture, carbonic acid, oxygen, and ammonia, and of retaining them so as to give them out to the roots of plants as they are wanted. This is done by mixing with it clay, which has these properties in a very considerable degree.

It not unfrequently happens that an abundance of clay is to be found lying underneath the sand at no great distance below the surface. When this is the case, clay is to be dug up and allowed to remain in small ridges, so as to be exposed to the sunlight, the air, the rain, and the cold of winter. After having been so exposed, for a year or longer, it is ready to be scattered upon the surface of the sandy land, or to be ploughed into it. The good qualities of the land will thus be permanently improved. It will be able to absorb and will become retentive of moisture, carbonic acid, and ammonia, and of all the manures. Such an addition may be called an *amendment.*

293. Another, and, after the clay, a still more effectual way of rendering a sandy soil fertile, is the application of large quantities of marsh mud, peat or swamp muck. There are often, in the immediate neighborhood of sandy fields, old mud holes, bogs, or swamps, where vegetable substance,—humus,—has been accumulating for centuries. This, by itself, is of no value. But when spread upon the land, and acted upon by the atmosphere, it immediately begins to act upon the silicates.

"The very act of exposure of this swamp muck has caused an evolution of carbonic acid gas. *That* decomposes the silicates of potash in the sand; that potash converts the insoluble into soluble manure, and lo! a crop. That growing crop adds its power to the geine."

By such processes, repeated from year to year, "it is not to be doubted, that every inch of every sandy knoll, on every farm, may be changed into a soil, in thirteen years, of half that number of inches of good mould."*

* Dana's Muck Manual.

And if this can be done with the barren sandy knolls, how much more with the plains!

294. Where neither clay nor marsh mud is to be easily obtained, light, sandy land may sometimes be rendered capable of absorbing and retaining the atmospheric elements of vegetable food and thus becoming fertile, by scattering plaster upon it and sowing clover seed. When the crop of clover, together with the weeds which will spring up with it, is in perfection, that is, nearly or quite ripe, it may be ploughed in. This process, though seemingly a waste of good clover hay, is one by which many poor lands may be rendered fertile and afterwards kept so by careful cultivation.

295. If it be objected that all these amendments require a good deal of time and labor, it may be answered, that there are days in the year when a farmer can spare both, and that a permanent improvement of land is worth a good deal of both. *There are no gains without pains.* Clay may be brought from a clay pit or muck from a bog at seasons of the year when no agricultural operation can go on.

296. A **Clayey Soil** is to be improved first by the application of sand, as fine as can be found, in quantities proportioned to the hardness and closeness of the clay. The object is to bring it into such a state as shall allow water to penetrate freely, and that it shall harden and crack less under the influence of drought. If applied to the surface, the sand will exert at once a favorable influence there, and will soon find its way down into the clay, when another layer may be applied. This may be done as well in the heart of winter as at any other season. The sand not only improves the texture of the soil, but the reciprocal action of the clay and the sand, aided as it

will be by any manure that may be applied and by the vital power of the growing plants, supplies new materials for their food.

A clayey soil is always greatly improved by deep draining.

297. A limestone or **Calcareous Soil,** in which there is a deficiency of sand or of clay, may be amended by the application of each, according to the means within reach. A valuable addition to a calcareous soil is the sandy mud found in the bed of a stream, which may often be easily obtained in the dryest part of summer.

298. A fourth kind of soil, naturally unproductive of valuable plants, is that of marshes and swamps. Unproductive as such soils are, they are mines of vegetable wealth, as they always contain an abundance of substance produced by the decay of vegetable and animal matters, —of the richest humus.

They are to be wisely husbanded. They often contain, in a single acre, enough of the organic elements of fertility to convert forty acres of hungry, barren land into fertile soil. This mine should not be covered over and lost, as it often is, by burying it under a coat of sand. If a farmer has many acres of swamp or marsh, he may bring a portion of it into immediate fertility by an exchange with the dry and sandy hills of the neighborhood,—a load of sand for the surface of the swamp for a load of muck for the surface of the hill,—but he ought to leave always a part of his mine accessible, at every season of the year, and continue to draw from it as long as he has an acre of poor sandy land left.

299. The soil formed from the swamp, by draining and covering with sand, may be greatly benefited by the application of lime, guano and other heating manures.

300. Soils in which clay predominates are usually *heavy*, *stiff*, *wet* and *cold*, and difficult to cultivate. But, when well drained, amended by the application of sand and of humus, and carefully tilled, they produce abundantly, and repay the pains and expense which have been bestowed.

Wet lands are cold because of the continual evaporation of the water at the surface. Every one knows that when a wet hand or face is exposed to the wind, it feels cool. As the moisture is converted into vapor, it takes up heat, and gives to the surface a sensation of coolness. In the same way evaporation renders the surface of a wet soil constantly cool.

301. But lands commonly dry are on that account warm. Sandy land retains heat far better than clayey or peaty land.

Color also has an important influence. Dark-colored soils absorb heat, while light colored soils readily reflect it. Most manures are dark-colored. Rich soils, therefore, naturally absorb heat, and rich sandy soils retain it, better than poor ones.

That color has an effect upon the power of absorbing heat is proved by Dr. Franklin's experiment. Place black, blue, red and white pieces of cloth on the snow in the sunshine, and, after some hours, the sun's heat will have been so abundantly absorbed by the black, that it will have sunk into the snow before the white has begun to grow warm, while the red will be just beginning to sink and the blue will have sunk almost as far as the black.

302. There are few places in this part of the country where the soil has been formed by the crumbling of the rocks just beneath the surface. In most parts of the

Northern and Middle States, the soil is made up, in a considerable degree, sometimes wholly, of sands or clays, drifted from the north. These are often called diluvial soils, from a belief, once in vogue, that they had been brought to the places where they are found by the action of a deluge (*diluvium.*)

303. When the native forests are cut down, and the land cleared of the undergrowth, and broken up by the plough, the soil is almost uniformly found to be fertile. In most parts of America, this virgin soil will bear large crops of grain and other valuable plants, for many years in succession, without manure. This fertility is owing to the fact that the surface has been occupied by forest trees and other forest plants for countless centuries. By the decay of the leaves, fruit, roots and trunks, the ground has been covered with a coat of humus or forest mould; and by weathering,—the long continued action of the atmosphere, and other great agencies of nature,—the minerals in the soil have been brought into a state suitable for the food of plants.

304. To give some instances of this action. The oxygen of the air, combining with the iron or oxide of iron in a particle of granite, makes it swell and crumble, and, at the same time, releases the potash or other element which had been associated with the iron, and leaves it ready to be taken up by the roots of a plant. Carbonic acid acts in a similar way upon lime and magnesia.

305. But the carbonic acid does not act alone. Carbonic acid is always ready to be dissolved or absorbed by water; and water, thus charged with it, has not only the power of dissolving limestone and magnesian rocks, but exerts a slow but certain influence by which even granite and the other hardest rocks are gradually crumbled; very

few minerals, perhaps none, being able to resist its long-continued action; and though its solvent power seems to be slight, in the lapse of time it produces changes of great importance and extent.

306. Carbonic acid acts in other ways. It unites with the ammonia of the atmosphere, forming carbonate of ammonia, and with the potash and soda in the earth, forming carbonates of potash and of soda. These three alkaline carbonates have the power of dissolving silica. Now it has just been stated that silica enters as an ingredient into the composition of nearly all the harder rocks. Of the three minerals of which granite is composed, quartz is almost pure silica; mica is two thirds silica, and felspar is about one half made up of silica. All these minerals and many others are thus gradually disintegrated by the slow action of these carbonates upon the silica in them.

307. Why does the fertility cease? The mineral and atmospheric elements of the food of plants are gradually taken up by successive crops, and carried off with them, the humus grows thin and meagre, and the soil is exhausted. The crops obtained from the land are, year after year, continually smaller, till at last they are not sufficient to reward the labors of the husbandman.

308. The obvious remedy is to restore to the soil the elements wanting, as will be shown in the chapter upon manures.

309. But if a soil be barren for one plant, it is not necessarily so for every other. A field which, for want of soluble silica, will not bear a second crop of Indian corn, may, from having a plenty of potash and lime in it, bear an excellent crop of clover or of beets or carrots. There may not be enough of a particular element for one

kind of plant, while a plant of another kind may find a quantity of food amply sufficient for its perfect development. A third sort of plant may thrive upon the same soil, after the second, if the remaining mineral constituents are sufficient for a crop of it. And if, during the cultivation of these crops, a new quantity of the substance wanting for the first, for instance, of soluble silica for Indian corn, has been rendered available by weathering, then, if the other elements be found in sufficient quantity, the first crop may be again grown upon the same land.

CHAPTER X.

OF THE SUBSOIL.

310. Immediately below the soil lies the subsoil. It may be and often is composed of the same kind of earth as the proper soil; or it may be entirely different. A sandy soil may rest upon a subsoil of clay, or upon calcareous rock, or rock of any other kind, or upon gravel.

311. The influence of the subsoil upon vegetation is often very great, especially when the soil is not deep enough for the free growth of the roots of the plants cultivated. In that case, when the subsoil is of such a nature as to admit of it, the soil should be deepened by ploughing. This should be done gradually and with judgment, because, as the subsoil has no mould or loam in it, turning too much of it up to the surface at once,

will be very likely to render the soil poorer for some time, instead of richer. If a farmer is aware that his soil would be improved by being deeper, he must make the improvement by adding to its depth a little each year.

312. When a loose sand rests upon clay; or a clayey soil upon calcareous marl, or upon sand; indeed whenever the subsoil will serve as an amendment to the soil, the two may be mixed with great advantage.

The evils of a subsoil impermeable to water are the stagnation of water and the excessive humidity of the soil. Generally, a very slight declivity is sufficient to induce the water to trickle along below the soil upon the surface of the subsoil, until it finds some means of escape. But even in this case, there is likely to remain in the soil superfluous moisture, which ought to be carried away by draining.

313. When the slope is not sufficient to lead the water to run off, the ground becomes boggy and the evil is declared by signs intelligible to every-body, by the springing up of rushes, sedges and other bog plants. But when the slope allows the water to trickle away slowly, the evil is not so apparent. The most certain sign, perhaps, is the presence of the weeds called horsetail, and scouring rush, (species of *equisetum,*) which need a subsoil always wet for their horizontal roots to run upon.

It may be laid down as a rule that wherever horsetail appears, the ground needs draining.

CHAPTER XI.

OF AMENDMENTS.

314. The soil plays, in the life of plants, a double part. It serves to give room and foothold to the roots; and it furnishes or keeps in store for plants the elements necessary for their nourishment.

The qualities a soil ought to have, to give sufficient foothold, must vary with the plants. The grains need a somewhat compact soil to give firmness of foothold; the different kinds of clover a deeper one. On the whole, what is best suited to plants is average qualities, a soil neither too compact nor too mellow, neither too heavy nor too light, too wet nor too dry.

315. These evils are remedied by **Amendments,** that is, operations, or the use of substances, by which the soil will be improved in its physical qualities. For example, increasing the humidity of dry soils, diminishing that of moist soils, increasing the tenacity of light soils, lessening that of heavy soils, or any other changes in the mechanical or physical properties, would properly be called amendments.

316. Argillaceous soils may be improved by the addition, not only of sand, but of gravel, broken brick and plaster, in short by any thing which will render them more open, loose and penetrable by air and water. In England, clayey land is often much improved by burning over the surface, or by burning a portion of the clay and scattering it upon the land. By burning, the clay changes its properties and becomes more like sand, and in this state loosens the soil.

317. The amendments suited to light, dry, siliceous lands, are clay, as already suggested, to give them cohesion, and argillaceous marls, whenever they are to be had. Irrigation not only gives moisture to a dry soil, but always brings useful additions in the substances which have been dissolved in the water and are deposited when the water is at rest.

318. Planting with trees, especially planting dry, barren hills with forest trees, permanently increases the moisture, not only of the surface covered by the trees, but of the neighborhood, and thus improves the climate. Draining is a valuable amendment.

319. In reference to a proposed amendment, the expense must be calculated, and the question must be settled whether the increased produce will pay for the outlay.

When the materials are near at hand and it will cost little to get them and transport them, the question is easily settled.

320. The character of the amendment must also be considered. A sandy soil amended by the addition of clay becomes permanently better. The clay can never be exhausted, and will always give to the soil the power of absorbing and retaining the elements of the food of plants.

An amendment produced by the introduction of humus or any form of carbonaceous matter will give value to the land, as long as it continues to be well cultivated and manured, but, like manure, the added matter is liable to be exhausted.

The quantity to be used will vary with the depth of ploughing.

CHAPTER XII.

OF FERTILIZERS.

321. The soil ought to contain all the *elements* necessary to the nourishment of plants. These have already been spoken of in the chapter upon the various elementary substances found in plants. They are: 1, oxygen; 2, carbon, in the state of carbonic acid; 3, hydrogen; 4, nitrogen, in the shape of ammonia; 5, silicon; 6, sulphur, and 7, phosphorus; 8, chlorine, and 9, sodium, in the shape of common salt; 10, calcium; 11, potassium; 12, magnesium; 13, iron; 14, manganesium. It must also contain 15, aluminum, as the basis of clay, and, though in minute quantity, 16, fluorine; and the water or the soil must contain for certain marine plants, 17, iodine, and 18, bromine.

322. These, except the first four, atmospheric elements, are always found in combination, as silicates, sulphates, nitrates, phosphates, carbonates and fluates, of potash, soda, lime, magnesia, iron, manganese and alumina, or in other forms sometimes more complex.

We know that these are all essential to plants, because we find them all in the ashes of plants.

If any one of these elements were absolutely wanting in a soil, the plants to which that element was essential could do little more than sprout there; and if planted or sown in such a soil, would starve to death. Plaster, for example, is essential to clover; and clover seed, sown in a soil which contained no plaster, would not come up. If there were a very little plaster in the soil, the clover might come up, but would not flourish.

323. What is the remedy? Plainly it is, to add to the soil the element or elements wanting; that is, *to apply manure to the soil.*

324. It might naturally be thought that, inasmuch as the atmospheric elements are furnished continually by the atmosphere, it could not be necessary to supply the soil with substances intended to furnish them. But then it must be remembered that the atmospheric elements are furnished very slowly, and it is always desirable to hasten the processes of vegetation, in our short seasons. It is therefore reasonable, and the experience of all agriculturists, in all temperate countries, shows it to be wise, to provide an abundant supply of those substances which are full of these atmospheric elements, or which serve to attract them and keep them in reserve for the wants of the growing plants.

325. To the question, therefore, Is nothing ever to be supplied to the soil but the mineral elements which are wanting? the answer is to be given, whenever humus is not already abundant in the soil, it is to be supplied. For humus furnishes directly, and also indirectly, by the changes that are going on in it from the action of the oxygen of the atmosphere and the vital power of plants, the carbonic acid, ammonia and nitric acid which are just as essential as the mineral elements.

326. But how are wild plants supplied with humus? By a process vastly too slow to meet the wants of the husbandman. The roots and leaves of the plants that have died, decay and form humus for those which are to succeed. But the supply is usually very scanty, and wild plants have often a thin, meagre look, in comparison with those under cultivation; as, for example, the slender-rooted wild carrot, when compared with the carrot of the

garden. Prof. Nuttall, who brought to this country many beautiful wild plants from Oregon, often said that when he saw them in the gardens of those to whom he had sent them, he could hardly recognize them, so much had they been improved in size and vigor by cultivation.

327. But humus is slowly prepared by the wild plants themselves. The *lichen* which encrusts the surface of a rock has no humus to begin to live on. It seems to have the power of eating into the rock itself and of extracting thence the mineral elements it needs. From the air and the rain it gets carbonic acid and ammonia, and, when it dies, deposits on the rock a thin coat of humus fitted for the partial nourishment of other generations of lichens. These are succeeded, after many years, by plants somewhat more fleshy, like the mosses, and by the grasses and other slender, longer rooted plants; and these by plants still larger; till, in the slow process of time, substance enough is gathered to give foothold to shrubs, and finally to trees.

328. The trees of the forest, by their annual deposit of leaves and, from time to time, of fruits, and at last by the fall and decay of their trunks, prepare a deep bed of humus or forest mould for the use of the husbandman.

Whenever he can, he avails himself of this treasure. But where it is wanting or scanty, cultivated plants are to be furnished with the abundant humus which they need, by placing in the soil, within reach of their roots, organic, that is to say, vegetable and animal substances, in the state of decay.

329. How these act has already been shown. They possess themselves and impart to the soil the power of absorbing and retaining, for the use of plants, the water and with it the carbonic acid, ammonia, oxygen, nitric acid

and other elements which come down dissolved in the rain. These, acting on each other, and quickened in their action by the air, by the sun's light and heat, and by the electric and vital influences of the plants, continually prepare for the use of plants, the food which they need, in the form best suited to their nourishment.

330. To the question, Which are more important, the atmospheric elements thus furnished, or the earthy or mineral? we answer, Both are equally important. Both are indispensable. They are necessary to each other. A soil rich in organic substances, attracts and retains the atmospheric elements in abundance proportioned to its richness. Such a soil puts the earthy elements into a condition suited to the wants of vegetation; and, the more readily and abundantly, in proportion to the fulness of the supply of these earthy elements.

331. **Fertilizers** may accordingly be divided into two great classes, viz.: **Inorganic** or **Mineral Fertilizers,** and **Organic,** or **Vegetable** and **Animal Manures.**

OF INORGANIC OR MINERAL FERTILIZERS.

332. In their general character, inorganic fertilizers are both manures and amendments. They furnish nourishment to plants, at the same time that they exert a mechanical action upon the texture of the soil, upon its lightness, stiffness, compactness, &c.

333. The principal mineral fertilizers are lime, marl, plaster, wood ashes, ley, soot, sulphates and other salts of ammonia, phosphates and super-phosphates of lime, common salt, carbonates, nitrates, silicates of potash and soda, sulphates of soda, of lime, and of magnesia, &c. But all of these are not in common use.

334. **Quicklime** is limestone, chalk, or shells, deprived of their carbonic acid by heat in a fire or a lime-kiln. Quicklime amends a soil by decomposing some of its ingredients, and by setting at liberty the potash and other alkalies which exist in combination with clay and in particles of granitic sand. It also hastens the decay of organic substances, and combines with some of the gaseous products given out during the process. It should be in a state of powder, before it is scattered upon the soil. It combines with the carbonic acid which is always in the air and constantly brought down by rain, and thus returns to the state of carbonate of lime.

This, by itself, is insoluble in water, but water containing carbonic acid has the power of dissolving carbonate of lime, and thus the carbonate so formed and that already in the limestone rocks are dissolved, and the rocks are disintegrated.

It also acts upon plants by diminishing the evaporation from their surface, and thus husbands the moisture in the soil, and makes it last longer than it would without the lime. This same effect is also produced by gypsum, nitre, common salt, and most of the other saline manures.

335. An excellent way of using lime is in a compost, as is practiced in Flanders. Make a layer of lime, and cover it with a layer of sods, weeds, scrapings of ditches and roads, river mud, marsh mud, and any thing else rich in organic substances. Follow with successive layers of lime and of the organic matter, and cover with a coat of loam. At the end of a fortnight, it may be worked over, and this may be repeated, from time to time. The longer it remains in a heap, the more complete is the mixture, and the better the compost.

336. Lime mellows clayey land. It is an essential element in most plants and is valuable therefore for itself. It is a very important element in tobacco, potatoes, pease, the clovers, and turnips. It corrects the acidity of soils, particularly of that of bogs and swamps. An examination of the mineral ingredients of our soils shows that it is never wanting.

337. Yet, in most parts of New England, it is so difficult to obtain and so dear that it cannot often be largely applied. In small quantities, it produces, when needed, most important effects. In England, large quantities are often applied to land in the shape of chalk.

338. Limestone rocks often contain magnesia, which is acted upon in a lime-kiln just as lime is. This diminishes the value of the lime, as does the mixture of clay and of sand, with which it is sometimes adulterated. Wherever oyster shells or any other shells can be readily got, they may be burned on heaps of brush, or other fuel of little value, and will be converted into a lime which is of greater value for agricultural purposes, than that formed from limestone rocks, because it contains a small quantity of phosphoric acid. The having already formed a part of an organized being seems also to prepare it for a similar service.

339. **Marl** is a mixture of lime and clay, or lime and sand, sometimes, but not often, found in New England, but abundant in some other States. When exposed to the atmosphere, it should crumble easily, as its action is in proportion to its readiness to mix perfectly with the soil. Though less energetic, it has all the permanent effects of lime, and is very valuable as an amendment, clayey marl to sandy soils, and sandy marl to clayey.

340. Plaster, or plaster of Paris, as it is often called, is *sulphate of lime:* and the valuable effects it produces upon soils are owing to its supplying them not only with lime, but with the very important and often essential element of sulphur.

341. Sulphur, or brimstone, is present in nearly all parts of vegetables and of animals. Mustard seeds and the seeds of all other *cruciferous* plants contain a large proportion of sulphur. It also exists in the white of eggs, in the curd of milk, in hair and in wool.

Several very valuable salts are formed by *sulphuric acid* or oil of vitriol. By combining with potash, it forms sulphate of potash; with soda, sulphate of soda,—Glauber's salt; with lime, sulphate of lime,—plaster or gypsum; with magnesia, sulphate of magnesia,—Epsom salts; with alumina, sulphate of alumina; with oxide of iron, sulphate of iron,—copperas. And it is from these and other similar compounds that plants derive the sulphur found in them.

342. **Plaster** produces a striking effect upon the water in which it is dissolved, "such water, being incapable of cooking vegetables and of dissolving soap, is called *hard water;* but it may be very easily and economically converted into *soft water*, and rendered fit for domestic and culinary purposes, by adding to it a small quantity of ordinary carbonate of soda, in the proportion of about half an ounce per gallon."—*Normandy.*

Carbonate of lime is formed, which settles to the bottom as a white sediment, from which soft water may be poured off.

343. Plaster has also the property of being decomposed by the carbonate of ammonia. It is thus turned into sulphate of ammonia, which is not volatile at a common

temperature, and so husbands the ammonia for the future use of plants. This takes place because ammonia and sulphuric acid have a greater mutual attraction than ammonia and carbonic acid. The ammonia, therefore, leaves the carbonic acid with which it has been united, and unites with sulphuric acid, to form sulphate of ammonia; and the lime, deprived of the sulphuric acid, unites with carbonic acid, to form carbonate of lime. This is more clearly shown by the following diagram:—

SULPHATE OF LIME. { Sulphuric Acid, ... Sulphate of Ammonia.
{ Lime, ...

CARBONATE OF AMMONIA. { Ammonia, ...
{ Carbonic Acid, ... Carbonate of Lime.

344. The carbonate of ammonia comes from the air, in which it is formed by the combination of the carbonic acid always floating there, with the ammonia always forming by the union of hydrogon and nitrogen. Or it may be formed in the earth.

345. But when and how should plaster be applied? When a soil does not contain naturally any sulphate of lime, or when it has been exhausted by cropping, the addition of that substance may prove of great value in two ways; 1st, by furnishing food for the plants mentioned, and 2d, by fixing the ammonia of the atmosphere and laying it up in store for the future use of plants by decomposing, as shown above, the carbonate of ammonia contained in rain water, and making soluble sulphate of ammonia and carbonate of lime.

When applied, plaster should be scattered, in the shape of the finest, impalpable powder, in the spring, just as vegetation is beginning, while the dew of the morning or evening is on the plants, that it may stick, but not in rainy weather.

346. The other sulphates are also useful. Sulphate of soda is said to produce good effects upon clover and other green crops. And so also is sulphate of magnesia good for these crops and for potatoes.

347. **Ashes.** In Westphalia there is a proverb that "he pays double who buys no ashes." It is a fact often observed that, on strewing wood ashes on a meadow which has long been mown, thousands of clover plants make their appearance, where none were visible before.

Ashes are made up of salts, such as silicates, phosphates, sulphates and carbonates. The carbonates and sulphates of potash and soda, as found in ashes, are soluble and are dissolved out by leaching. The silicates, phosphates and carbonates of lime, magnesia, iron and manganese, are insoluble and thus remain in leached ashes. A portion also of silicate of potash remains undissolved.

Far the larger part of leached ashes is carbonate of lime. The next is phosphate of lime or bone dust.

248. Unleached wood ashes are of great value in the cultivation of many crops, especially Indian corn, turnips, beets and potatoes, because of the great amount of carbonate and other salts of potash which they contain, and so important is potash to these plants that they are often called potash plants.

349. Leached ashes are of less general value, but still are a very valuable fertilizer, by reason of the salts which they contain, which, though not soluble in simple water, may be rendered soluble by the influence of other salts, of air, and of the vital power of plants, and may be thus again taken up into the circulation, and again perform

the service they had already performed in the plants from the combustion of which they came. They have important effects when mixed in compost heaps.

350. The ashes of sea coal and anthracite are not without value, and have a good effect upon cold, stiff soils, and are found an excellent top-dressing for grass, even on light soils. As they absorb water and the gases, they are deodorizers, and retain the offensive gases for the food of plants. They have a slow but good effect, scattered among trees, and are particularly valuable in the formation of walks and roads.

351. Since ashes lose some of their good qualities by having ley drawn from them by leaching, **Ley** itself must be useful as a manure; and not only ley, but that which is left after the ley has been made into soap by combining with fats and oils, and done its office as soap by taking dirt from clothes, dishes, faces and hands. Soap suds and dish water, therefore, are so valuable that they ought never to be lost or thrown away. They have an excellent effect if sprinkled upon grass or other growing crops, or poured upon compost heaps.

352. **Soot** is a precious manure, since it is made up of carbon, in the state of the finest powder, and is full of volatile salts. In Flanders, it is reserved for beds of colza, which it protects against plant lice. In England, it is scattered upon meadows, where it promotes the vegetation of grass, while it destroys moss. Three large crops of clover have been got in one year by the use of it. The soot from bituminous coal is still better than that from wood.

353. As **Carbonate of Potash** and **Carbonate of Soda** are the forms in which potash and soda are found in ashes,

they must have the same effects as ashes, only in a more decided manner.

354. The salts of ammonia, especially the nitrate, are very valuable as manures, and are particularly applicable to soils already rich in phosphates, or which contain vegetable acids. **Sulphate of Ammonia,** which may be obtained at a moderate price at the manufactories of gas, is excellent, when applied in small quantities, to fields of meadow hay or of wheat.

355. **Nitrate of Potash,** East India saltpetre, is nitric acid and potash united. As might be expected, both nitrate of potash and **Nitrate of Soda,** South American saltpetre, yielding not only nitrogen but potash and soda to plants, are particularly beneficial to wheat and to barley.

356. And, as the plants grown in the fields must supply the phosphate of lime which is essential to the growth of the bones of all animals, and this ingredient in soil is likely to be exhausted, **Phosphate and Super-phosphate of Lime** are of the very greatest value as manures. Phosphate of lime is usually applied in the shape of ground bones, and super-phosphate, as bones dissolved by sulphuric acid and diluted with water, applied either in a liquid state, or reduced to powder by drying.

357. All the elements in the salts of ammonia, of potash and of lime, here spoken of, are either taken up by plants, or exert a most important influence upon the humus in the soil, hastening the process of decay, and converting insoluble into soluble salts.

358. Common salt is also sometimes of great value as a fertilizer. For some plants, asparagus, for example, it is of indisputable importance, and may be employed in very large quantities. It not only enriches the soil for

asparagus, but it kills nearly all the weeds; and as weeds are commonly nothing but valuable plants out of place, it must be used with discretion, or it may do more harm than good.

Applied in small quantities, it has the effect of rendering grass and clover more pleasant to animals, and, in a small proportion, it is of the greatest value to all cultivated crops. It is also a valuable addition to the farmyard and to the compost heap. Salt which has been used in curing fish or meats is much cheaper and far better than pure salt.

359. The object of manures is to give to the soil whatever is wholly or partly wanting to it, whether of a combustible or an incombustible nature. The use of organic manures is to furnish the soil with humus, geine or mould, which shall serve as a reservoir, to hold in readiness, for the use of plants, all the kinds of food necessary to their growth. And the use of humus is to furnish and keep a ready supply of carbonic acid, ammonia and water, which three are the last result of the decomposition of vegetable substances.

360. Such being the object, organic manures should be employed in a condition favorable to decomposition, either in a fermented state or, better still, ready to enter into fermentation. Manures which should refuse to decompose would be of no use. But the decomposition must not be too far advanced. Ammonia is very volatile, as its common name indicates, and may readily escape into the air and be lost. The penetrating, characteristic odor of ammonia is perceived in stables, near manure heaps, and wherever else nitrogenous substances, that is, vegetable and animal substances containing nitrogen, are in a state of decay. Every one who has had occasion to use

a smelling bottle, knows the effect of ammonia upon the organs of smell.

When the manure is immediately covered up, the ammonia, as it is disengaged, is kept in the soil, especially if there be clay or loam or something else present which has an attraction for it.

361. **Organic Manures** are divided into Vegetable Manures, Animal Manures, and Mixtures of Vegetable and Animal.

The principal vegetable manures are green crops, kelp and rock-weeds, straw, sedge or reeds, leaves, brewer's grains, &c.

362. **Green Manures** are standing crops, ploughed in, if possible, when ripe, for it is then that they contain the greatest quantity of soluble matter. The best plants for the purpose are the different kinds of clover, lucerne and sainfoin, vetches, buckwheat, cabbage-leaves, radishes, turnip-tops, wild mustard and wild turnip, potato-tops, Indian corn, rye, &c. Yet some of these are better suited to certain soils than others.

To be suited to this purpose, plants should grow rapidly, so as not to occupy the land too long; their seed should be cheap, and they should be plants which borrow most of their elements from the atmosphere. Such plants bestow upon the soil more than they receive from it.

363. The green crops best suited to light and sandy soils are buckwheat, the clovers, cabbages, radishes, wild mustard, potato and turnip-tops, rye, and Indian corn. For stiff, clayey soils, beans and pease, the different kinds of clover, vetches, &c. But green crops are less suited to clayey than to any other kind of soil. For calcareous soils they are exceedingly advantageous, as such soils

need no lime. For all other soils, especially clayey soils, lime should be scattered profusely upon the green crop at the time it is ploughed in. On very dry, sandy soils, the use of green manures is very beneficial, as they speedily decay in such soils and supply vegetable mould, which, being retentive of water, does something to correct the want of such soils and is very serviceable in time of drought.

364. Green manuring is particularly applicable to mountainous districts, and those remote from the homestead, where the expense of carriage of other manures would be too considerable, and also to poor soils deficient in clay, and which, on that account, imperfectly retain water.

365. For winter wheat, or winter rye, to both of which green manures are well suited, the land should be ploughed deep in spring, and the seed for a green crop be sown so that it shall be ripe a week or two before the winter grain is to be sown. The green crop sown with lime or plaster should be ploughed in to a moderate depth, say two to four inches, and, just as the decomposition is beginning, the wheat or rye should be sown. The grain, as it sprouts, and while it is young, will thus take advantage of the ammonia and other products of the vegetable decay.

366. Where land is very much infested with weeds, two green crops may be grown, the same season, and ploughed in before the weeds are ripe. Most of the seeds of the early and also of the late weeds will thus be made to come up, and the plants be turned in, with the green crop, for the benefit of the soil.

367. The addition to the soil is not the only advantage of green manures. The mechanical condition of the ground is remarkably altered by the ploughing in of plants and their remains. A tenacious soil thereby loses

its cohesion; it becomes more friable and more readily pulverized than by the most careful ploughing. In a sandy soil, coherence may be given. Each stem, of the green plants ploughed in, opens, by its decay, a road by which the delicate rootlets of the future plant may ramify in all directions to seek their food.

368. Kelp and rock-weed are very valuable as a manure. They contain a good deal of nitrogen and a large proportion of alkaline and earthy salts, and, as they undergo decomposition more rapidly than other green manures, so their effect upon vegetation is, proportionally, much more powerful, but it is also much less lasting. The slender, grass-like sea-weed, also called eel-grass, has very little value as a manure, as it has little substance, and yields very slowly to decay, but is still valuable for its mechanical effects upon heavy soils.

Kelp and rock-weed may be ploughed in, like other green manures, but this should be done as soon as possible, or, if this is not practicable, they should be stratified with earth and lime, in order to convert them into a compost, or they may be mixed with ordinary manure.

These sea-weeds act beneficially on all ordinary crops. If spread upon grass in spring or early summer, they promote its growth; and a crop of grain subsequently obtained from such a soil, is said to be much improved, at least in quantity, for the quality is thought to be deteriorated. In the north of Scotland, farmers prefer kelp and rock-weed to any other manure for cabbages. They form an excellent manure for flax and hemp, the flax obtained being improved thereby, both in quantity and quality. Rye, oats, turnips and clover are benefited by that manure. Their action upon vegetables is immediate but does not last long, showing its effects, however, more the second year than the first.

369. The straw and leaves of particular vegetables are the best manure for those vegetables, wheat straw for wheat, potato-tops for potatoes, and the leaves and prunings of grape vines for those vines.

Straw ploughed into stiff clay soils renders them more porous and thus lets in the air, and causes decay not only of the straw but of the organic matter previously existing there. Wheat and other grain stubble on stiff soils should be ploughed in soon after the grain is removed, both for the reason just given, in regard to straw, and because, the fresher the roots, the more rapidly do they decompose. This does not hold true for light sandy land.

For hay land, or land to be laid down to grass, damaged hay, not fit for animals, is valuable as a manure. Sedge and the reed-grass of salt marshes are also of use, but less valuable than the substances just mentioned.

370. The leaves from different trees have very different degrees of value. Poplar leaves, oak leaves and chestnut, beech, and maple leaves, are rich in nutritive matters, while thinner leaves and pine leaves contain very little nourishment for plants. The leaves of the larch are considered favorable to grasses, from the fact that hills planted with larches afford better pasturage than they had furnished when they were bare. But this may be the consequence of the land being shaded. All leaves should be ploughed in as soon as possible after they have fallen. Leaves, grasses, young twigs, and all other green vegetable matter, the very element of humus, are valuable as manures, and their value is greater in proportion to their freshness when ploughed in; and whatever is valuable in this way is valuable for the compost heap.

371. **Animal Manures.** They are more powerful than vegetable or mixed manures, on account of the great quantity of nitrogen which they contain, and the important salts which exist in them. The nitrogen unites with hydrogen, and forms ammonia, and this the ammoniacal salts. These dissolve other mineral substances, and are absorbed by water, which carries them down to the roots of plants. The more abundant these elements of plant food are, the more rapidly will they enter into plants, and the surer and more abundant will be the crops. The more completely the soil has been mixed and pulverized, the more readily will the roots reach their supply of food.

372. The flesh of quadrupeds, fishes and other dead animals, contains about 50 per cent. of carbon, and from 13 to 17 of nitrogen, besides water, salts of potash and soda, of lime and of magnesia, and is therefore one of the very best manures that can be.

These substances, and all offal and animal refuse, should never be applied directly to the soil, but made into a compost.

373. The best way of disposing of the carcass of a dead animal is to place it in a hole one or two feet deep, sprinkle an abundance of quick-lime upon it, then throw on a layer of earth, then a layer of plaster, then a layer of earth mixed with powdered copperas, and then a sufficient depth of earth. The plaster and copperas absorb the ammonia and sulphuretted hydrogen, as they are formed, and prevent all unpleasant effluvia.

In a few weeks, the heap may be opened, the bones separated, to be used in bone manure, and the remaining mass turned over and mixed, if necessary, with additional earth. This, repeated once or twice, will make the substance ready for use. (*Normandy.*) The body of a dead horse can convert twenty tons of peat into a manure richer and more lasting than stable manure.—*Dana.*

374. Sulphuretted hydrogen is a nauseously smelling compound of sulphur and hydrogen. It gives its peculiar smell to a rotten egg. When dead fish or fish offal is thrown upon land, it not only diffuses a most offensive smell to a great distance, but it imparts a very disagreeable flavor to the crops, and also to the milk and to the butter made from the milk of cows who feed upon such crops.

375. Hoofs, hair, feathers, skins, wool, and blood, contain more than 50 per cent. of carbon, and from 13 to 18 of nitrogen, besides sulphur, and salts of lime, of soda and of magnesia. They therefore hold the first rank among manures, and, as a long time is required for their decomposition, their action may last for seven or eight years. They yield excellent results, made into a compost for potatoes, turnips, or hops, or for meadow land.

376. Hair, spread upon meadows, augments the crop threefold; and, the Chinese, who know its value, collect it every time they have their head shaved,—and the operation is performed once a fortnight,—and sell it to the farmers. The crop of hair, from the head of each individual, amounts, in a year, to about half a pound. Every million of persons therefore affords two hundred and fifty tons of hair, that is, of manure of the most valuable kind, since it represents at least two thousand five hundred tons of ordinary barnyard manure, and which might be collected without trouble, but which is now invariably lost. You may calculate what must be the loss for the State, and for the whole United States.

377. Blood, besides more than 52 per cent. of carbon and 17 per cent. of nitrogen, contains soluble salts, such

as common salt, phosphates, sulphates and carbonates of potash, soda, &c., water, and some insoluble salts, namely phosphate of lime and of magnesia. Like flesh, it should be made into a compost with other substances, and it thus becomes a very valuable manure for light soils, while its effect on clayey soils is less obvious.

378. Bones contain more than 53 per cent. of phosphate of lime, a little phosphate of magnesia, some carbonate of soda, &c., and more than 7 per cent. of nitrogen. Their principal value is owing to the quantity of the phosphates they contain, as these salts are largely removed from a soil by the feeding of cattle and by successive crops. These salts remain after the bones have been deprived of their fatty substance by the soap-boiler, though most of the nitrogen is lost. Bones should be ground before being used, and may be applied at the rate of ten or twelve hundred weight to the acre. Even when ground, they produce effects which may be seen for several years.

379. The action of bones may be accelerated by converting their phosphates into perphosphates or superphosphates, which is done by mixing the ground bones with half their weight of sulphuric acid diluted with three or four times its bulk of water. This is to be thoroughly mixed and left a day or two at rest. One barrel of the pasty mass may then be mixed with one hundred barrels of water and sprinkled upon the field from a water-cart or by scoops. Or the perphosphate may be mixed with a large quantity of earth, or sawdust, soot or powdered charcoal, and thus applied to the land.

380. It is easy to see how it comes that animal manures are so valuable. Animals live almost wholly upon substances derived from the vegetable kingdom. These substances, restored to the earth, from which and from the

air they must originally have come, are naturally, there fore, the very most important elements of the food of plants.

381. **Mixed Manures.** It is the uniform experience of all farmers and gardeners in all parts of the world, that barn manure, that which comes from the stable, the cow-house, the sheep-fold, the pig-sty and other similar sources, is, on the whole, the most valuable and the most universal in its beneficial effects of all known manures. Other manures have great value for particular purposes. This is useful for all. It is the only manure which keeps up the fertility of all kinds of land. This is just what we should expect. Many plants are cultivated as food for cattle and other animals. The concentrated essence of the nutritious elements of plants goes to form the bodies of animals; and we have just seen how extremely valuable as manure, is every part of those bodies. A portion is converted into milk. We know how precious, primarily as food and indirectly as furnishing butter and cheese, the milk of cows is. In the mountains of Europe, and among the poorer classes, the milk of goats and of sheep, is not less precious. In the great plains of Arabia and Tartary, the same priceless advantages are afforded by the milk of the camel and the mare.

All these valuable elements of vegetable food, except the comparatively small portion which is converted into flesh or milk, are or should be thrown upon the manure heap.

382. Manure is of such primary importance upon every farm, and there is so much danger that valuable portions of it should be washed away by rain and lost in the earth, or dried up by the sun, or wafted away by the winds, that particular care should be taken to secure it.

The best and most convenient arrangement, when it can be made, is to have the manure fall into a cellar immediately under the stable or cow-house. And care should be taken that no portion, liquid or solid, should be lost. If it be left exposed to the open air, and suffered to be drenched by rain, or parched up by the sun, a great quantity of the products of its decomposition will be volatilized or washed away. There is danger also of its heating, from the process of decomposition which immediately begins, especially in the cellar under the stable for horses. The temperature should not be permitted to exceed 100° of Fahrenheit, and if a smell of ammonia be perceived, it is a proof that the valuable products of its decomposition are wasting; and means must be immediately employed to fix them, that is, make them combine with something else, and thus prevent their loss.

383. This can be done by watering the manure heap with dilute sulphuric acid, or a solution of copperas, (sulphate of iron,) or by sprinkling plaster over it, when the odor of ammonia will immediately disappear. In a cellar, however, where the liquid manure is as carefully saved as the solid, and into which a stream of water may be directed by a spout from the gutter under the eaves, there will seldom be danger of heating, and a little fresh garden soil or loam thrown in may produce all the most important effects of the chemical substances.

384. By **Decomposition** is meant a change among the elements of a compound substance and their union in other forms. This takes place in consequence of the attraction which the elements have for the oxygen of the air and of water. The vital principle counteracts this attraction. In an egg, for example, as long as there is life in it, the contents remain unchanged and are ready to be waked up into a living creature. But as soon as the life is gone, decomposition begins, the sulphur and

hydrogen in the egg, warmed a little, attract each other and form sulphuretted hydrogen, which is ready to fly off, and oxygen unites with the other ingredients, forming new compounds.

385. **Fermentation.** The oxygen of the air is always ready to unite with other elements. If the juices of plants containing sugar, such as cider, or wine, for example, be carefully kept from the air, they remain sweet. But if the air be admitted, the oxygen immediately unites with the albumen of the juice, and then with the sugar, and the **Vinous Fermentation** begins. If this is allowed to continue, the sugar will be changed into carbonic acid and alcohol.

Weak wine, cider or beer, exposed to air, at the temperature of from 70° to 90°, gradually grows warmer, and becomes thick by slender threads moving in every direction through it, with a low hissing noise. When the noise has ceased, and the threads have attached themselves to the sides and bottom of the vessel, the liquor, now become clear, has passed through the **Acetous Fermentation,** and become acetic acid or vinegar.

386. The final products of complete decay are universally the same. The carbon of organic bodies combines with oxygen and forms carbonic acid. The hydrogen unites with oxygen and forms water, or with nitrogen and forms ammonia; or with sulphur and phosphorus, forming sulphuretted and phosphuretted hydrogen. The incombustible matters alone remain. Moisture and warmth are necessary at the beginning and at every stage of decomposition. To prevent it, therefore, we have only to keep the substance cold and dry.

387. It is desirable to keep the stable and cow-house always clean and sweet; and this may be effectually done by sprinkling a little plaster upon the floor once a day.

We commonly think that a stable or a cow-house is necessarily a dirty place. Why ought it to be kept clean and sweet? It is almost quite as essential to the health and comfort of horses or of cows, that they should be kept clean and allowed to breathe a pure atmosphere, as it is for the health and comfort of human beings. Besides, cows are often milked in their stalls; and if so penetrating a substance as ammonia fill the air there, it will necessarily be absorbed by the milk and give it a bad taste and smell.

The cost of a little plaster is very trifling. Enough to answer the purpose for a whole winter will not cost a dollar; and the value of the manure will be increased far more than that, so that you have only to pay a little pains for the pleasure of being clean and having the animals clean, with a sweet smelling place for them to live in and yourself to go to.

388. The products of the stable, of the cow-house, of the sheep-fold and of the pig-sty, are not of quite the same composition and value. They are different and suited to different uses. As a general rule, the contents of the cellar under the cows and oxen are more fit for very dry, light soils, and those from the horse-stable for stiff, clayey soils. The scrapings of the sheep-fold are better suited to meadow lands, as they often impart a disagreeable flavor to culinary vegetables; and the same is true of the contents of the pig-sty.

The common practice of throwing every kind of manure into one cellar, to form one heap, is not a bad one.

When the soil to be cultivated is an average soil, neither a stiff clay nor a dry sand, but a free, arable soil, the practice is a very good one. The defects of one kind of manure are corrected by the qualities of another, and such mixed manure will be neither too cold nor liable to heat and burn. It is of manures of this kind that the French proverbs have been made: "A small manure heap never fills a large corn bin." "It is not he that sows but he that manures well that gets the crop." "Less seed and more manure." "Without manure there are no good fields; with plenty of manure there are no poor ones."

389. The best materials for litter or bedding for cattle and horses are straw of every kind, damaged hay, sedge, reeds, leaves, sawdust. If these cannot conveniently be had, turf may be used, or loam, or even sand, which has the advantage of keeping animals free from lice. It should be something which will help to make them warm in cold weather, and dry and clean at all times. Horses and cattle should be always kept nicely clean. Both look better, fare better and fatten better, when they are carefully curried or carded and rubbed every day.

It is an excellent plan to have the cellar floor of clay firmly rammed and made even, but sloping towards the middle from the sides, and from the middle towards one end. There, in a place easily reached, should be a hollow to receive the liquid from the heap. The manure will be greatly benefited and prevented from heating, by pouring this liquid, from time to time, upon the top of the manure heap. Or, if the heap does not need it, it may be poured, with great advantage, upon compost heaps. Flemish manure is a liquid manure formed in a cistern, to which drains from the bottom of vaults bring the most valuable of all manures. Into this cistern water is made to run, which completely dissolves and dilutes whatever is in the vault. The liquid is sprinkled by a watering cart over meadows and growing crops, with striking effects.

390. A valuable liquid manure is formed by mixing with the liquid from the manure heap any other rich substances with a large quantity of water, which is to be poured by means of the sprinkling cart upon growing crops.

It can be applied, advantageously, to those fields which are already rich enough in humus or mould, as one great benefit of the application of manures in a solid form is to furnish a permanent reservoir for moisture, carbonic acid, and ammonia, and other elements of the food of plants capable of being dissolved in water or of being absorbed by decayed vegetables, and kept ready for the use of plants.

391. What is the most valuable of all manures, the statement of a few facts will enable you to judge. The principal object in view in the cultivation of all cereal plants, all leguminous vegetables, all fruits and nearly all roots, is, directly or indirectly, to furnish food for man. Most of the animals which he has domesticated, the sheep, the ox, the swine, all kinds of domestic fowls, the birds shot by the fowler and the fishes caught by the fisherman, are intended to supply his table. Now, of all these substances, vegetable, fish, flesh and fowl, which enter into the human system as food or as drink, for the supply of man's wants or as luxuries, all, except the little which is used to build up and to renew his body, is thrown away and is usually lost.

392. Chemical analysis entirely confirms the conclusions of common sense in this matter. The body itself, as is well known, is continually changing; its substance is becoming effete and its elements are constantly renewed. Chemical analysis shows that all the substances which have been enumerated as the elements of plants, all the

gases, the carbon, sulphur and phosphorus, all the alkalies and all the earths and metals, are not only found in the substance of the different parts of the human body, in the bones, the brain, the flesh, the tendons, the skin, and the delicate humors between them, but they are all found in those substances which have formed a part of the human body or have been within it, and have been cast out, after having performed their necessary and beneficent offices.

Now all these substances, literally the concentrated essence of soils, of vegetable and of animal organization, are usually thrown away and lost. If restored to the soil, they would more effectually renew it, and restore its fertility than all other manures and amendments put together, and yet they are allowed to escape and to be utterly wasted. And not only are they wasted and lost. Substances which, if properly preserved and husbanded, would render fertile as a garden the neighborhood of all great towns and cities, and would keep up the fertility of all the farms throughout the country, are now allowed to flow away into drains and sewers and to poison the atmosphere of towns and the waters of the rivers. There is scarcely any other instance of so enormous a waste. Chiefly in consequence of this waste, the farms, in all the older parts of the country, are becoming, or are already become, far less productive than they originally were. Even in those parts of New York and of the West that have been longest settled, though all recently settled, the fields are already losing their fertility from the same cause.

393. What means ought to be employed to prevent this waste? Economy, as well as regard for cleanliness and health, demands that measures should everywhere be taken to save all these substances, of every kind, liquid

and solid, to mix them with such substances as will render them inoffensive, and afterwards to compost them with other materials for manure and to restore them to the soil.

394. Many substances will prevent all disagreeable effluvia; plaster, copperas, Glauber's salt, sulphuric acid, or, better still, Epsom salts, chloride of manganese, sulphate and chloride of zinc, chloride of lime, all of which substances can be procured at a very low price. Most of these are completely soluble in water. Plaster is not so, and should therefore be put into those places only which are regularly and thoroughly cleared out.

395. If the above mentioned substances and all others capable of being used as manure, were always carefully husbanded and used, there would be no necessity for the use of guano.

Guano, (pronounced gooahno,) is the Peruvian name for the droppings of sea-fowls, found upon certain uninhabited islands on the coast of Peru and of Africa, in a climate not subject to rain. Guano has been accumulating there for an unknown length of time. It is found in deposits of great depth and is now dug out and exported to Europe and the United States, as a substitute for or an adjunct to farm yard manure. Guano consists principally of alkaline and earthy phosphates, and of ammonia and ammoniacal salts or compounds capable of being resolved into ammonia.

Good guano, exposed to a heat of 212°, loses not more than from 6 to 12 per cent. including a little ammonia. Poor guano, or that which is in a state of advanced decomposition, loses as much as 35 or even 40 per cent. of water.

396. Is it not very discouraging that after all the pains a farmer takes to fill his soil with valuable manure, it should be all washed away or into the deep earth by the rain? It would be very discouraging if it were true, but fortunately it is not true; as is made very apparent by a simple experiment or two. If a funnel be filled with soil, and a dilute solution of silicate of potash be poured upon it, there will not be found in the filtered water, as it runs out of the funnel, a trace of potash, and, only under certain circumstances, silicic acid.

If a funnel be filled with earth, and water, holding in solution ammonia, potash, phosphoric acid and silicic acid, be poured into it, none of these substances will be found in the water escaping from the funnel. The soil will have completely withdrawn them and incorporated them with itself.

397. Or make another experiment. Take a portion of garden soil full of potash, silicic acid, ammonia, or phosphoric acid, put it into a funnel and pour water upon it. The water will not dissolve out a trace of it. The most continuous rain cannot remove from a field, except mechanically, that is, unless it carry off soil and all, any of the essential constituents of its fertility. It is a common fear that the nourishing substances in liquid manure and in guano, will, if not immediately taken up by plants, be lost. But the fear is wholly unfounded. From liquid manure diluted with much water, or from a solution of guano, soil, when used in sufficient quantity, removes the whole of the ammonia, potash, and phosphoric acid which they contain. Not a trace of these substances can be found in the water which flows from the soil.

398. It is probable that plants sometimes obtain mineral elements which they need from the rocks themselves; and there are some facts which make it certain that they do so. We frequently find, in meadows, smooth lime-stones with their surfaces covered with a net work of small furrows; and we find that each furrow corresponds to a rootlet, which appears as if it had eaten into the stone. So, lichens grow upon the surface of bare rocks; and forest trees form vast trunks, full of potash and other salts, on the rocky soils of hills from which all the loose soil has been washed. It seems probable that their rootlets have the power of decomposing the rock and taking potash from the felspar or mica they find in them.—*Liebig*.

399. Is it necessary that each particular element of plants should be present in the soil? Or, if one be wanting, cannot plants be sustained by the others?

The agriculturist requires eight substances in his soil, that all the plants may flourish luxuriently, and his fields produce the largest crops. These eight substances are like eight links of a chain round a wheel. If one is weak, the chain is soon broken, and the missing link is always the most important, without which the machine cannot be put in motion by the wheel. The strength of the chain depends on the weakest of the links.—*Liebig*.

Those eight are phosphoric acid, potash, silicic acid, sulphuric acid, lime, magnesia, iron, and chloride of sodium. All these are essential to the growth of plants. Still more essential are oxygen, hydrogen, nitrogen and carbon; but these are always supplied by the atmosphere, in the form of water, ammonia and carbonic acid.

400. If we cannot obtain stable manure or other animal manure, how is the want to be supplied? Chemists

know exactly what substances are contained in stable manure, and they are able to point out artificial manures which contain all these substances and may be used instead of stable manure; and the most important of these have already been pointed out under the head of inorganic fertilizers.

401. **Composts.** How is the stable manure to be husbanded so as to go as far as possible? One way is by the proper management of the compost heap. Loads of marsh mud, of swamp muck, of earth from bogs and the bottom of ponds and rivers, are to be thrown into the manure cellar or upon the compost heap. The manure heaps and the compost heaps are to be turned over and over, till the contents are thoroughly mixed.

So great is the value of muck or swamp mud, for this purpose, that a farm is hardly to be considered complete without a swamp, or muck hole. Fresh turf forms a very valuable addition to the manure cellar and compost heap. This may be taken from the sides of roads and of walls and fences.

Peat taken from the sea side, where it has been daily covered with sea water, and mixed with one seventh its bulk of slacked lime, heats and ferments, and produces excellent effects as a manure. Any peat, saturated with strong brine, and mixed with lime, would be equally effective.—*Dana.*

Every farmer should make his own compost heaps, according to the materials he has for them, always taking care that no vegetable or animal substance be allowed to be lost.

Mud from the bottom of lakes, ponds or pools, is always of much value as a material for composts, especially when it has been long lying there. In every piece of still water, many animal and vegetable substances will have collected and been completely decomposed. The mud at the bottom will be made up of the remains of these substances and of earth completely saturated with their elements. Such mud must be full of fertilizing material. It is therefore a great and unnecessary waste to allow the scourings of hills near the homestead, and especially of streets and roads, to pour themselves directly into brooks and rivers, and to run off and be lost in the sea. A little care may prevent this. They may be made to pour upon low grounds, and a low mound of earth may detain them and allow them to deposit their mud.

402. A compost for trees to be planted on meagre, sandy soil, should be prepared of clay well mixed with muck or marsh mud, and with lime or marl. For clayey soil one of sandy loam, light muck and lime, with a portion of barn manure. Bog earth or peat, with lime, makes a good compost for almost any land except boggy land. To each of these a most important addition is ashes, or potash, or substances containing potash. The leaves of all trees, indeed all leaves, and weeds, and the small branches of all shrubs, are rich in potash, and are a natural manure for trees.

These, prepared long before hand, and thoroughly mixed with the soil, will have a surprising effect upon the growth of trees.

A good compost for any common crop is made of one cord of barnyard manure, with two or three of muck, swamp mud, or loam, and ashes or potash.

A compost which has been successfully tried by a careful observer is made of farmyard manure, twenty-five bushels, muck or mud, twenty-five bushels, and six bushels of leached ashes, or, in place of the ashes, one bushel of lime slaked with salt water.

A practical farmer of great experience and judgment, says that a good compost for hoed crops is formed of thirty bushels of swamp muck thoroughly mixed with one of guano.

Another excellent compost, recommended by the same person, may be made of the same quantity of muck with two bushels of good bones.

Another; dig peat or swamp mud, in the fall. In the spring, mix eight bushels of ashes with every cord; or, with every cord, twenty pounds of soda ash, or thirty of potash, dissolved and poured carefully upon the pile.

403. **Care in the management of the Manure Cellar and the Compost Heap essential to the health of the farmer's family.**

We have seen that ammonia, sulphuretted hydrogen and other gases should not be lost, as they are valuable as elements of the food of plants. But there are other and still higher reasons why such gases should be carefully prevented from coming out into the air.

These gases, while they give life to plants, are death to men. Sulphuretted hydrogen is not only very disagreeable to the smell, but it is thought, by some persons who have carefully investigated, so poisonous that, if it float in the air breathed by human beings, even in the proportion of one part to 100,000, it sometimes causes death. In one case, "a strong, healthy man came home from his work and went to bed. An hour had hardly elapsed when he was found dead." In another instance, a healthy child was taken ill in the morning and was a corpse at night. In both cases, the air breathed was analyzed and found to contain sulphuretted hydrogen. If breathed, even in very small quantities, it produces stupor, or causes a low fever, which, if the sufferer be not relieved by removal to perfectly pure air, may end fatally.* Carbonic acid when breathed in the proportion of 15 to 20 parts in 1,000 of air, causes immediate distress and feelings of suffocation, accompanied often with giddiness and headache. This is sometimes followed by a slight delirium and then by an irresistible desire to sleep.* If breathed in still larger quantities it not unfrequently causes death. The fumes of smoking charcoal, in a close room, have often been fatal to people sleeping in the room.

404. The *effects*, if breathed *in smaller proportions*, are dulness, heaviness, difficulty of thought, and apparent stupidity. The extreme sleepiness and dulness sometimes observed in children who have remained several hours in an ill-ventilated school-room, are, doubtless, often caused by the carbonic acid in the air of the room.

This comes from the breath of the occupants of the room, and sometimes from the fire-place or stove. Ammonia, breathed when very strong, immediately takes away the breath. When weaker, it irritates the lungs, and, even when very weak, if breathed for a considerable time, it produces symptoms of typhoid fever.

405. These poisonous gases are generated in drains and sink-holes, in heaps of dirt of any kind, in damp cellars and close rooms, in dirty ditches, in muddy puddles, swamps and undrained marshes, and wherever water is allowed to remain stagnant.

These poisons show their presence by rendering the air disagreeable to the sense of smell. Whatever is offensive to this sense is more or less dangerous; and, if foul air, that is, bad smelling, fœtid air, be breathed, it is always

* Dr. Taylor, as quoted bv Dr. John Bell. Third National Sanitary Convention, p. 425.

more or less poisonous. The poison may act slowly, but not the less surely, and it renders a person who breathes it liable to fever, cholera, consumption and other fearful diseases. It is universally found that people living in damp and dirty places, in houses ill-ventilated, over wet cellars or on ground badly drained, are the first to be attacked by cholera, dysentery, and various kinds of fever.

406. What has this to do with agriculture? Much. It shows that the farmer who looks everywhere for manure, and collects it carefully from all dirty places, of all kinds, secures his own health, and improves the health and comfort of his family and of his neighbors, at the same time that he improves his fields and increases his crops. The sweepings of rooms, the scrapings of cellars, earth that has been long lying under barns or other buildings having no cellars, the contents of drains, cess-pools, ditches, bogs, dirty ponds, morasses and swamps, are all excellent materials for the compost heap. Collected together and covered with clay or loam, they become not only harmless but very valuable.

All kinds of dirt, if allowed to remain near dwelling-houses, are liable to be dissolved or rendered noisome by the rain, and to sink into the earth and reach and contaminate the water in the well. Water thus contaminated is not only nauseous to the smell and to the taste, but very unwholesome. On this account the compost heap should always be made at a distance from the well; and beneath every such heap there should be an abundance of clay or loam, sufficient to absorb all the valuable substance that drains from the heap, and to prevent the moisture from sinking into the earth.

CHAPTER XIII.

OF TILLAGE.

407. In what does the preparation of soils consist? In various operations, the object of which is to divide and mellow the soil, in order to render it permeable to air, to water, and to the roots of cultivated plants, and so to mingle all the parts of the soil that all the elements of the nourishment of plants may be so diffused as to be within the reach of the roots, and also to keep it clean and free from weeds.

When the land is wet, the first and most indispensable of operations is draining. The essential operations afterwards are ploughing, digging, spading, harrowing and rolling.

Ploughing is turning over the soil, so as to bring a lower portion to the surface and to place in contact with the subsoil the portion which had been previously exposed to the air.

408. The objects of ploughing are to mellow and pulverize the soil, to mix it, when necessary, with a portion of the subsoil, to mingle the different portions as fully as possible, to cover manures, to destroy weeds, and to keep the surface fresh. All these things except the two last, can be done more effectually with the spade, the shovel and the fork, than with the plough. Weeds can often be better destroyed and the surface be more easily kept fresh by the horse-hoe or the cultivator.

Why then is the plough preferred? Because it is so great a labor-saver. The ground may be more easily and better turned over, in long slices, and placed upside down, by the plough, than by any other instrument which has been contrived.

409. What is the object in bringing fresh portions of earth to the surface? Soils have a remarkable property of attracting moisture from the air and condensing it in their pores. With the moisture, they at the same time absorb the ammonia, nitric acid and carbonic acid floating in the air or dissolved in the water. By long contact of the soil with the air the surface hardens and acts less efficiently, and the pores become filled. Hence the advantage of bringing a new portion into action.

410. **Deep Ploughing** extends all the benefits of tillage to a greater depth. It opens a larger portion of the soil to the beneficial action of the air and moisture, and affords a larger space for the food laid up for the use of plants. It distributes the manure more evenly through the soil. It has the effects, already mentioned, (Art. 47,) of draining. It gives you more land to the acre,—a new farm under the old one. Soil deeply ploughed is less speedily exhausted. The roots penetrate deeper and take firmer hold. Grain sown on deep soil is less liable to lodge.

If the food for plants is mixed evenly throughout the soil to the depth of ten or twelve inches, the roots of most cultivated plants will penetrate to that depth in search of it; and will thus be less liable to injury from drought.

411. Deep ploughing produces a saving of labor as well as of land. If a farmer who has commonly ploughed his field six inches deep, will plough, the present year, to the depth of seven inches, and will put on seven loads of manure where he had previously put on six, he will, with the same labor, get seven bushels of roots or of corn, where he has commonly got only six. If then, the

next year, he will plough eight inches deep, instead of seven, and apply eight loads of manure, instead of seven, he will find his crops increased in that proportion, upon the same land and with no more labor. The next year, or at the beginning of the next rotation, he may, on the same principle, plough to the depth of nine or ten inches.

It is only in this gradual way that the change can be safely made. And at each deepening, care must be taken to have a sufficient portion of manure put into that part of the earth which is last brought to the surface, in order that the plants while young may be made to throw out a great number of rootlets. This number will depend upon the amount of manure near the surface, in the immediate neighborhood of the little plant. These rootlets, once formed, will penetrate into the deeper earth and feed upon the food there prepared for them.

When the soil is *too rich* in carbonaceous matter, burning over the surface, and thus reducing bushes and weeds to ashes, is a very useful operation. We commonly get potash, which is so valuable to all vegetables, from the ashes of wood; but the ashes of shrubs and of herbaceous plants contain more potash than the ashes of the same weight of timber.

LAND not sufficiently RICH in vegetable remains SHOULD NEVER BE BURNT OVER.

412. USE OF THOROUGH TILLAGE. The more completely the particles of a soil are reduced to powder, the more readily they act on each other; and the more evenly the manure is diffused through the soil, the more readily and immediately do the roots of plants come in contact with them and feed on them. The only difference to be found between some very rich soils in Ohio and some very poor, was the fact that, in the rich soils, the same mineral con-

stituents were in the state of the finest powder. All mineral substances combine with oxygen and with each other the more readily in proportion as they are reduced to more minute particles.

413. Most people are wholly unaware of the value of tillage. As a general rule, we may say, the more completely and frequently the soil is stirred the better. Farmers are apt to think that the great advantage of hoeing and cultivating with the plough, the harrow and the cultivator, between rows of corn or other crop, is the destruction of weeds. This doubtless is indispensable. But in reality, the improvement of the soil by continually exposing fresh portions to the air, by thoroughly mixing it, and thus preparing for future crops, is of not less value than the weeding. Though, doubtless, there may be danger of too frequently turning dry soils in a season of drought.

414. Subsoiling is cultivating with a plough which does not turn a furrow, but penetrates to some distance below the furrow already turned and loosens the soil down there. It sometimes adds one third to the crop raised. By stirring and loosening the earth to a considerable depth, it makes it retentive of moisture to that depth, and, with moisture, of all that accompanies moisture into the earth, and makes it easy for the roots to penetrate and reach them.

If the roots of a plant do not penetrate so deeply, their food, deep in the earth, reaches them by capillary attraction. This draws the moisture, and all that the moisture contains, up towards the surface. A part of it is taken up by the plants, and the remainder, as the moisture evaporates, is left near the surface to be still farther acted upon by the air.

CHAPTER XIV.

PREPARATION OF LANDS.

415. A texture or mechanical condition of the soil favorable to plant growth is especially necessary. The mechanical condition of the soil is its condition in respect to looseness or compactness, hardness or mellowness, coarseness or fineness, without reference to the chemical substances contained in it.

416. Few soils are naturally in the mechanical condition best suited for cultivation, though different soils vary very much in this respect. Hence it is as necessary to use the right means to put the soil into the proper mechanical condition, as to apply manure to improve the land in the other modes above referred to.

417. The soil must be mellow, so that the roots of plants can penetrate freely and the air can circulate through it, but still firm enough to hold the roots in their position. It must admit the heat of the sun, and yet hold moisture enough for the wants of the plant.

418. Most soils require to be well pulverized before they allow the roots of plants to penetrate and grow freely, or permit the circulation of the atmospheric air, and if they are not so pulverized and mellow, they do not readily take up and carry off the water which falls in rain or comes from other sources. This water often washes away the surface of the soil, or remains stagnant, causing much injury to vegetation.

419. The manner in which land must be prepared for cultivation, differs very much in different cases, varying

according to the condition in which it is found when its improvement is first begun. The processes most frequently found necessary are clearing, draining, ploughing, harrowing and rolling.

420. Clearing is generally required in a new country, or when new land or woodland is to be cultivated. In these cases the soil rarely allows even the most ordinary operations of farming. It is often covered with trees or forests, or with rocks which would interfere very much with successful tillage.

421. The term clearing, in a new country, is applied to the cutting down and burning or removing of all the timber and brushwood from the lot. This is simple, though hard work. The trees are felled, if possible, in June, when in full leaf, and the ground may be burned over in season to sow in a crop of winter rye upon the surface. This is the case in remote sections where the timber has so little value as not to pay for removal, and where it is usually burned on the ground. But in other locations, the wood may be cut and removed in winter, and the work of clearing continued the following summer. Sometimes on account of its situation, the cleared land must be devoted to pasturage. In these cases grass seed is sown along with the rye, and cattle turned upon it the following season. But generally the sides of steep hills, or land so rough that it cannot be cleared and prepared for cultivation except at great expense, should be kept for woodland.

422. The next step in preparing wild lands for farming, is to remove the stumps and stones. Several simple machines have been constructed to do this, by which a powerful leverage or purchase is gained, so as to raise a stump or stone of several tons weight from its bed. A

convenient and cheap form of stump puller is Bates' patent, shown in figure 2, and one of the best forms of a stone lifter in figure 3.

Fig. 2.

Fig. 3.

423. It often happens that the surface is completely matted with roots of bushes, and so hard as to be impen-

etrable to the plough in pasture or waste lands which it is designed to clear up. In such cases a stout grapple represented in figure 4 is found extremely useful in removing the surface which may be burned previous to ploughing.

Fig. 4.

424. Much land is so situated as to require thorough draining before it can be cultivated at all to advantage. The object of draining is to remove an excess of moisture from the soil.

425. Water standing stagnant in the soil diminishes the good effects of manures very much by preventing decomposition, makes it impossible to work lands early in the spring, prevents seeds from germinating, or makes them germinate more slowly, and delays the ripening of crops, lessening their quantity and making their quality inferior.

426. An excess of water in the soil also excludes the air. This is injurious, because the air does much to promote the chemical changes in the mineral parts of the earth which are necessary to the growth of plants, and converts the organic materials in the soil into vegetable acids which give it the name of "sour" or "cold" soil.

427. Drainage is effected either by opening channels on the surface, or by means of covered drains. Open drains are sometimes very useful, but are liable to serious objections. The water which enters them, carries with it many of the substances which make the soil fertile, which are thus lost. Besides, such drains are not nearly as useful as covered ones, while they interfere with a proper cultivation; they leave a great deal of water in the soil, weeds are very apt to grow along their sides, and they

take up a great deal of ground which might otherwise be made productive.

428. Underdrains avoid these objections, and are more economical. They may be constructed either of stones or of tiles made for the purpose. The tiles are altogether better, both because they can be laid down at less expense, and because they last longer. They are also less liable to get stopped up.

429. To lay a stone drain properly, a large trench must be dug. This requires great labor, and such a drain should not be made unless there are a great many small stones on the surface of the land which the farmer wishes to get rid of, and even then the tile drain costs less and is more economical in most cases. The different modes of laying a stone drain are shown in figures 5, 6, and 7.

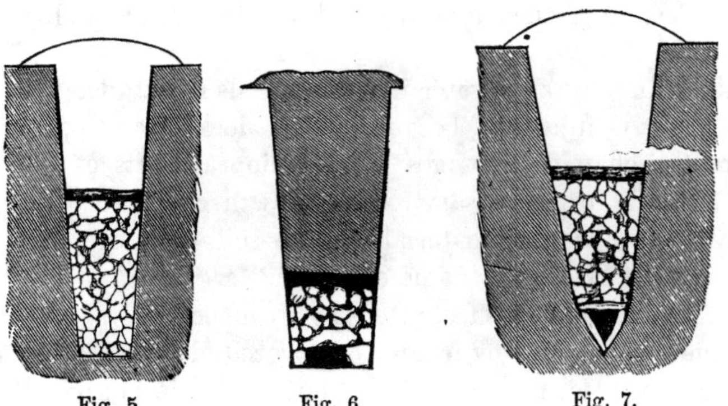

Fig. 5. Fig. 6. Fig. 7.

430. In laying down the tile drain, the trench may be very narrow, a width of a foot at the top and four inches at the bottom being sufficient, as in figure 8. It is dug by a spade and hoes made for the purpose, and illustrated in figures 9 and 10.

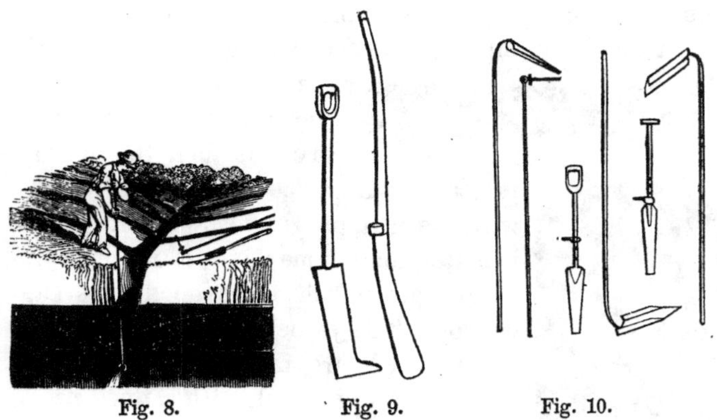

Fig. 8. Fig. 9. Fig. 10.

431. The tile drain is not only more economical, but it carries off the water better and lasts longer. If the stones were picked up and placed at the edge of the trench without any expense, the drain made of them would be less economical in the end than one made of tiles which cost $10 or $12 per thousand.

432. The pipe tile, (Fig. 11[a],) a simple round tube, is found to be the best in shape. For the interior drains which enter into the larger main drains, a tube of two inches in diameter is about the right size.

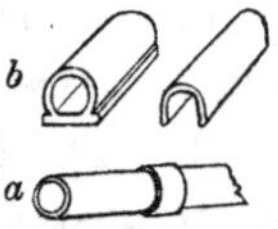

Fig. 11.

433. The fall should not be less than one inch to the rod. A drain properly laid in this way may be expected to last and answer a good purpose for half a century.

434. The sole tiles made in this country, shown in figure 11[b], are not so good because they must necessarily be laid sole down, and if they happen to be warped in burning, as they often are, it is difficult to get a perfectly straight and reliable water course.

435. The brush drain is sometimes made by digging a

trench and filling up to a certain depth with small brush. When this is attempted, the sticks should all be laid with the larger ends down, as shown in figure 12. The brush is then thoroughly pressed down and covered over with sods with the turf or grass side down. This is better than none; but it is never to be recommended where good tiles can be got. The same may be said of log drains which are made by laying down two logs in the trench with a third upon them, as in figure 13. The earth must be pressed down solid over a stone, brush or log drain.

Fig. 12.

Fig. 13.

436. The distance apart at which the drains should be laid will depend on the character of the soil. In a soil which is stiff and holds water long, it might not be well to have them more than twenty-five feet apart, while a more porous soil might be sufficiently drained if they were thirty or forty feet apart, or even more.

437. The depth of the trench must depend somewhat on the distance between the drains. Trenches three feet deep and twenty feet apart, have been found to do as well as those five feet deep and eighty feet apart. In general the depth should be from three to four feet.

438. Thorough draining makes the soil more open and causes a more free circulation of air through it, thus preventing it from drying up so soon. The air is at all times charged with moisture, and as it comes in contact with the particles of soil, this moisture is condensed and deposited there, just as we see it deposited on the cold sides of a pitcher of ice water in a hot day. Drainage

also deepens the arable soil and makes it more easy for plants to extend their roots.

439. The atmosphere is charged with fertilizing elements as well as with moisture, and as it circulates freely in the soil, these elements are taken up and retained to serve as plant food.

440. The soil having become more porous by the removal of water and the admission of air among its particles, its temperature is raised in consequence, that is the soil is made warmer and warmed to a greater depth.

441. A higher temperature in the soil hastens forward the growth of plants, and thus often makes the ripening several days earlier.

442. The texture or mechanical condition of most stiff soils is improved by simply draining, and they are thus made capable of being worked earlier in spring and after long rains, while the growth of plants is stronger and more vigorous. The difference may be seen in figures 14 and 15, the former showing the effect of draining and warming the surface soil, *a*, causing the roots to penetrate even into the moisture below the drained level at *b*, the latter, the same species of plant on an undrained and unsuitable soil.

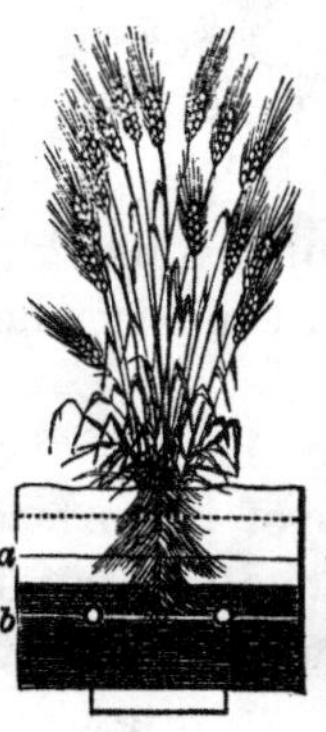

Fig. 14.

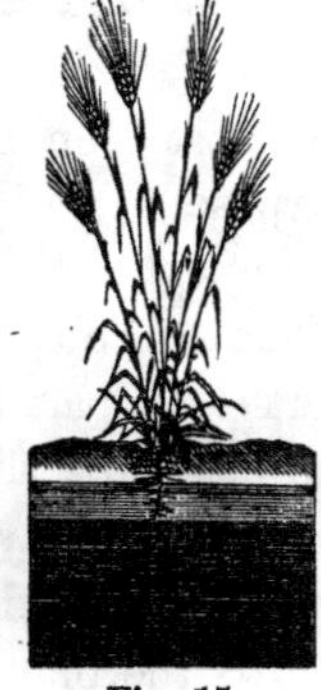
Fig. 15.

443. After the land is properly cleared, it must next be made ready for planting. In the first place the soil must be mellowed or broken up fine to a proper depth.

444. The spade, the plough, the harrow and the roller,

are the implements most often used in effecting this object.

445. The spade or spading fork is the simplest form of these implements, and consists of a blade or tines of iron or steel fixed into a straight handle. It is worked by hand. Cultivation by its use is the slowest and most expensive mode of tillage, and is adapted chiefly to the nice operations of the garden.

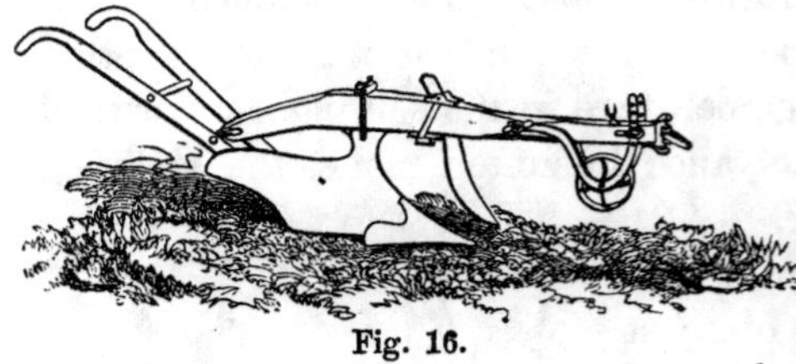

Fig. 16.

446. The common plough, (Fig. 16,) is the implement most commonly used in breaking up the land, and is the most economical instrument that can be used for the purpose. Without the plough successful farming would be impossible in a country where labor is very high and difficult to obtain.

447. In passing through the soil the plough separates and cuts off a slice of its surface, cutting it both vertically and horizontally, and turning it over in such a way as to leave it exposed to the action of the harrow, which usually follows the plough to break down and pulverize the soil completely.

448. The furrow made by the common plough should be deep, straight, and of such a width that the slice cut off may be turned entirely over, or left on its edge, as the ploughman may wish.

449. The depth is of the greatest importance, though experience has shown that it is best to deepen the arable soil gradually, by ploughing about an inch or half an inch deeper each time, till it is worked deep enough, say from

seven to ten or twelve inches, according to the crops it is designed to cultivate.

450. If much of a poor subsoil should be brought up to the surface at once, the farmer would have to wait two, three, or even four years before he would obtain the largest results, though after that time the good effects of deep tillage would be seen.

451. Deep ploughing has much the same effect as thorough draining, though in a less degree. It enables the roots of plants to penetrate deeply in search of nourishment, carries off more or less of the surface water, warms the soil, and without doubt makes it more fertile by allowing the air to circulate through it, and by a mixture of the soils of different depths. Besides, deep ploughing makes it much easier to do the other work which is necessary in preparing the soil for planting, and increases the effect of all manures which are applied.

452. Deep ploughing is especially needed in the cultivation of deep or tap-rooted plants like carrots, parsnips, and ruta-bagas, but it is beneficial to all crops if it is properly done.

453. The subsoil plough, (Figs. 17 and 18,) is designed to follow in the furrow of the common plough, to loosen and break up the lower layers of the soil without bringing them to the surface. With this implement it is easy to loosen the subsoil six or eight inches below the furrow left by the ordi-

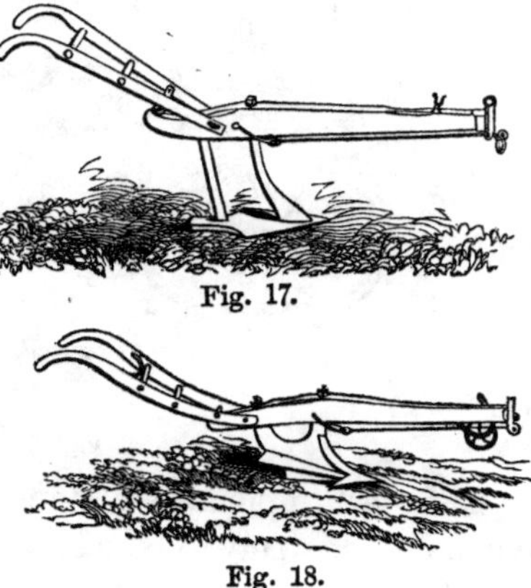

Fig. 17.

Fig. 18.

nary plough, making the whole depth to which the land is stirred, from eighteen to twenty-four inches.

454. The benefits of subsoil ploughing are very similar to those of deep ploughing. Recent investigations show that nitrogen and other fertilizing substances exist deep below the surface. Subsoil ploughing enables the roots of plants to reach them by loosening the soil to a greater depth.

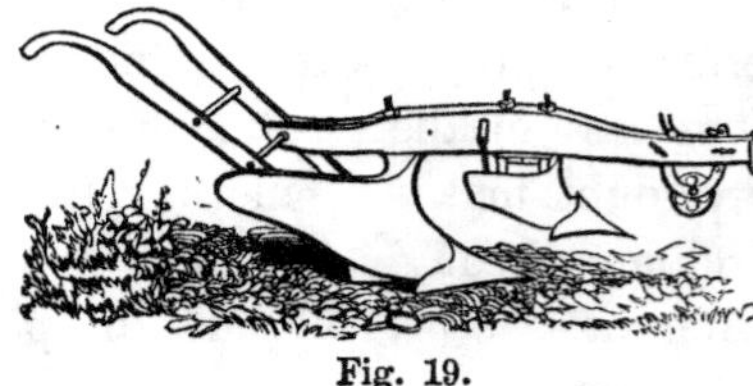
Fig. 19.

455. A very excellent implement known as the Michigan, or double mould-board plough, (Fig. 19,) is designed to obviate the necessity of the subsoil plough, to some extent. The smaller mould-board cuts off a thin surface and turns it into the last furrow, where it is completely covered with a finely pulverized soil by the principal mould-board.

Fig. 20.

456. An implement designed to supersede the use of the plough in many soils, is known as the digger, (Fig. 20.)

It leaves the ground mellow like the fork, and in good condition for the cultivation of crops.

457. The harrow, (Fig. 21,) is an ancient implement, and is most commonly used after the plough, to break down and mellow or pulverize the furrow slice. It should be moved rapidly over the soil. It has been very much improved within a few years.

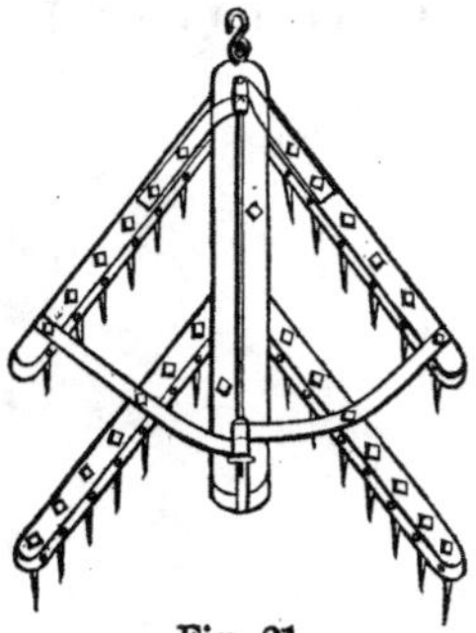

Fig. 21.

458. The cultivator, (Fig. 22,) may properly be regarded as a modified form of the harrow, but it is much better than the harrow, because with its plough shaped teeth, it lightens up and mellows the surface soil, instead of pressing it down hard, as the harrow is apt to do every where except on new, rough land.

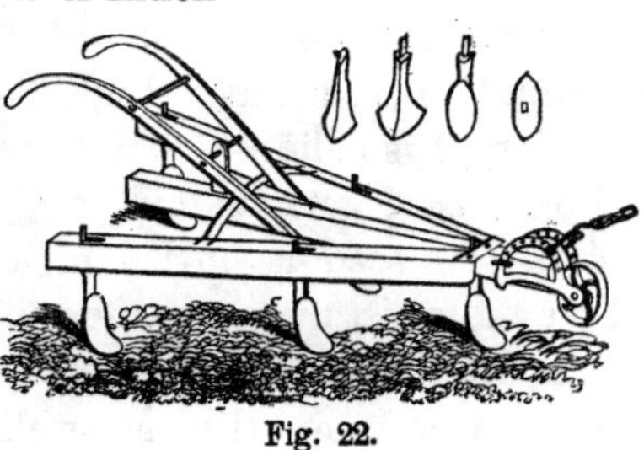
Fig. 22.

459. The roller is often used to pulverize the surface soil. It has so large a surface to rest on the soil, that it crushes and breaks up clods without hardening the lower strata.

460. In laying down lands to grass, it is often useful in pressing down small stones, so as to get them out of the way of the scythe. It is often useful, also, on newly sown grain, and hastens the germination of seeds, by preserving the moisture around them.

461. But clayey soils should never be rolled except when they are perfectly dry, and for the purpose of

breaking the lumps left by the plough. Rolling stiff soils when wet, would only make them too hard and compact, and thus do them more harm than good.

CHAPTER XV.

SOWING, PLANTING, ETC.

462. Moisture, warmth, and exposure to the air, to some extent, are needed to make the seeds of plants germinate healthfully. Light is not necessary; on the contrary, it is believed to interfere in some degree with the process of germination.

463. The seed is buried in a properly prepared soil, where the moisture soon softens it throughout, and certain chemical changes take place, by which the mealy parts are prepared to nourish the swelling germ.

464. A radical shoot or rootlet first bursts its covering, and invariably grows down, fixing itself in the soil, while a stalk shoots up towards the air and light in which it expands its leaves.

465. By means of its leaves, which serve as its lungs, the plant draws much nourishment from the air. There are a great many small openings or pores in the leaves, which are most numerous on the under side. On a single square inch of the leaf of the common lilac, there are no less than one hundred and twenty thousand of these little mouths, and on an inch of the white lily there are sixty thousand. They are found in great

numbers on the leaves of all plants. A magnified portion of the leaf of the grape is shown in figure 23.

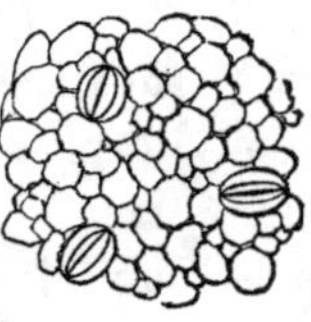

Fig. 23.

466. All plants come from seeds, in the first place, and the farmer usually sows or plants the seeds of the plants he wishes to have; but in some cases tubers or bulbs are placed in the ground and new plants spring from them. A tuber is a thickened portion of a stalk or stem under ground, having buds or eyes, as the potato and the artichoke. A bulb is a collection of fleshy scales formed under ground by certain kinds of plants, as the tulip, the onion, and the lily.

467. Generally the seeds are sown where the plant is to remain. But sometimes they are started in a carefully prepared seed-bed, from which they are transplanted to the field, where they can grow up to better advantage. This is done to bring them forward earlier.

468. For their complete development, all cultivated plants must have a deep, mellow soil, and care enough to prevent them from being injured by weeds or insects while they are growing. The farmer must also attend to the choice of seeds, taking only those which are good and still have the power of germination, and must consider how much seed he is to use, how he should prepare it, the time and manner of sowing, and the depth to which the seed should be covered.

469. **Choice of Seed.** An imperfect seed may still be capable of germination and may produce plants, which appear to grow well at first, but such plants will have a sickly and imperfect growth, especially at the time of flowering, and they will produce little grain and that of an inferior quality.

470. With the same soil, climate and cultivation, the most perfect seed will produce the finest crop. No seed is likely to produce a healthy and vigorous plant, unless it came from a strong and healthy plant itself, was fully ripened, and is so fresh that its power of germination is still uninjured.

471. Good seed may be known by its weight, its size, its glossy surface, and its freedom from any disagreeable odor. Plumpness and weight indicate that it was produced by a vigorous plant; a glossy covering shows it to be healthy, and the absence of odor shows that it has been well preserved.

472. To learn whether the germinating power still exists, we may take two pieces of thick cloth, moisten them with water, and place them one above the other in the bottom of a saucer. Then take some of the seeds, spread them out thin upon the cloths, not allowing them to cover or touch each other. Cover them over with a third cloth like the others, and moistened in the same manner. Set the saucer in a moderately warm place, and moisten the cloths from time to time, taking care not to use too much water. Good seed, thus treated, will swell gradually, while old or poor seed which has lost its germinating power, will become mouldy and begin to decay in a very few days.

473. Such a trial enables the farmer to judge whether old seed is mixed with new. The new germinates much more quickly than the old. It enables him, also, to judge of the quantity he must sow, since he can thus tell whether a half, three-quarters, or the whole will be likely to germinate, and will know what allowance to make for bad seed. Clover seeds, if new and fresh, will show their germs the third or fourth day.

474. The seeds of some plants continue good much longer than those of others. Those of many wild plants, for instance, will lie for many years without losing their goodness, if they happen to be in such a place that they cannot germinate, and afterwards when they have heat and moisture, and other conditions necessary for germination, they will produce plants.

475. In digging wells, or in other deep excavations, species of plants not before known in the place, often spring up from the earth thrown out. These seeds must have been lying in the earth many years, unable to grow because the heat and air could not reach them.

476. The seeds of the turnip, if kept in a dry, cool place, continue good several years, and will germinate nearly as well when five years old as when only one or two. But the seeds of the grasses are comparatively worthless when two years old, since few of them will then germinate. Age, heat, moisture and fermentation, are most injurious to seeds.

477. **Change of Seed.** Most of our cultivated plants originally grew wild, and in their natural state were much less valuable than they now are. They have been brought up to their present condition, and made far more useful for the nourishment of men and animals, by careful cultivation for many years. In all these plants there is a natural tendency to lose what they have gained, and fall back to their original condition. This can be prevented in some degree by constant care in the selection of seed and high cultivation; but experience shows that in some places these plants will gradually lose their best qualities, however much care may be used to guard against it.

478. To avoid the evils of sowing inferior seed, we may use that produced in other localities, where special care

is taken to raise it in the highest perfection and purity. In general, seeds should be preferred which were raised on a soil poorer than that where they are to be sown.

479. When both soil and climate are favorable, the necessity of frequent change may be avoided by good cultivation, and by taking the seeds from the best and most vigorous plants, when they are fully ripe, and drying and preserving them properly. Where this can be done without danger of deterioration, it is far better, since the farmer knows better what he is to sow. Where the species of plants cultivated are very similar to each other, and liable to hybridization or mixture, care must be taken to keep them so far separated as to preserve their purity.

480. The maxim that "Like produces like," so well known among farmers, may be true to some extent in regard to most of the cultivated plants of the farm, but we constantly see instances where the fruit of the plant which grows from a seed, is different from that of the plant which produced the seed sown; very common examples of this change are seen in the apple and other fruits, and the potato when raised from the seed. In our common cultivated grains, the difference, if there is any, is slight.

481. In a large field of wheat, a few specimens might be found among the millions of plants, which would differ from the seed planted. By carefully selecting these and planting them by themselves, new varieties may be obtained and preserved distinct.

482. So by taking care to select our seed corn from the ears which ripen earliest, we can get early varieties. If we choose seeds from the largest ears, and plant them by themselves, we shall obtain large varieties; and many persons think that if we take our seeds from those plants

which have several ears on a stalk, we may thus make very prolific varieties.

483. In these and similar cases, the change or modification from the original to the new variety is not generally sudden, and soon accomplished, but is most commonly slow and gradual. The seed must be carefully selected year after year, till the desired change is fixed and firmly established. New and somewhat permanent varieties may be thus obtained.

484. But the case is different when we cultivate potatoes and other tubers, since we do not usually plant the seed in such cases, the tubers being only an enlargement of the stem beneath the soil, and when plants grow from their buds or eyes,—as they do in the ordinary manner of raising potatoes, the same variety is extended or increased with no change of character.

485. New and distinct varieties of the potato may be produced to any extent by sowing the seeds of the plant. Thus the chenango, the pinkeye, and other varieties, were first obtained from seed taken from the ripe bolls of other varieties. After a new variety has been once made in that way, it may be continued and kept up by planting the tubers in the usual way.

486. If a vine is produced from a layer of another vine, the new vine is only a portion of the old one, and can never become a new and distinct variety; and so in budding or grafting, the new growth is only a portion of the same old tree from which the scion was taken, and has precisely the same character as the tree from which the bud or graft came, except so far as it may have been changed by the difference of soil or locality. But if the seeds of the apple or of the grape are sown, new varieties are obtained at once.

487. **Quantity of Seed.** The plants should cover the whole ground, each having just room enough to allow its full and complete development and no more. To learn how much seed will be necessary for this we must consider the character of the soil, its preparation, its fertility and the time of sowing. The quality of the seed, the extent to which it is apt to tiller or send up side shoots, and the manner of sowing must be taken into account; also the habits of growth of the plant—whether it is large and rank or otherwise, and the mode of tillage to be adopted—all these must be regarded.

488. The richer the soil and the more manure there is used, the ranker the plant will grow. The ranker the growth the more space will it require for its full development. On the other hand, in a poorer soil, the plant will grow less rankly, so that more seed will be required to cover the ground with plants on poor and scantily manured land than on rich land well manured.

489. The better the seed the less will be required. If the climate and soil are very favorable to the plant, a smaller quantity of seed will be needed, since a larger number of plants will grow from the same quantity of seed. So the earlier the sowing is finished, the less seed may be used provided the season is favorable.

490. If the soil is perfectly clean and free from weeds less seed is necessary. Much also depends on the distribution of it, and the more uniformly it is spread the less is required. For this reason hand or broad-cast sowing requires more seed than machine or drill sowing. In general, it may be said that winter wheat and rye, and other winter grains, require less seed than the spring varieties.

491. Other things being equal, thin sown crops ripen later than thick sown ones. The greater the space allowed each plant the more vigorous will be its development, and consequently, the slower its growth. In thick sown crops the growth is more quickly finished, and though the stalk may be rank the ear will be smaller, and the number of grains to a stalk less than in thin sown crops. By thick sowing we gain in time, but lose, to some extent, in quality.

492. **Steeping Seeds.** Some farmers are in the habit of soaking the seed in warm water, or in some solution like carbonate of ammonia, lime water, chloride of sodium or brine, partly to hasten its germination and partly to supply the place of manure. When the sowing has been delayed till after the proper time, this practice may be useful, but it is better to sow or plant at the right season, and so avoid the necessity of any thing of the sort to make the seed germinate more quickly, and as a substitute for manuring the land properly, this practice is of very little benefit.

493. The moisture of the soil is best adapted to nourish the germ, and the growth of the plant through the season will, generally, be more healthy without the use of any artificial preparation.

494. **Time of Planting.** The time of planting varies according to the season and the nature of the plant. Some grains, for instance, will endure a great degree of cold during the early period of their growth. It is generally considered better to sow these in autumn, and spring sowing would not do well. Others cannot bear much cold and should not be sown till spring. The condition of the soil, also, makes a great difference. A dry, warm soil is ready for planting much earlier in spring than a cold, clayey one.

495. The time of sowing should be suited to the nature of the particular plant we wish to cultivate. Indian corn, barley and buckwheat, for example, should be planted when the ground is dry and warmed by the heat of the sun, while certain kinds of wheat and oats do better when sown in a colder soil.

496. Winter grains should be sown earlier on heavy soils than on sandy ones, and earlier in a cool, moist climate than in a dry, warm one. There is no general rule as to the time of sowing which can be applied to all cases, and the farmer must always be governed by the circumstances of his own case.

497. **Depth of Covering.** The seed should be covered to such a depth as to secure the amount of heat, moisture and air, necessary for its germination. This depth varies with the kind of plant, the nature of the soil, the climate and the time of planting.

498. It is evident that on a clay soil which is less easily penetrated by air and warmth, the seed should be covered less deeply than on a sandy one. Spring planting ordinarily requires greater depth than autumn.

499. Very small seeds require only a shallow covering, and in many cases, a simple rolling without the use of the harrow, is sufficient to secure perfect germination. In common farm cultivation great losses often occur from covering seed too deeply, especially the smaller seeds, as those of the grasses and the clovers.

500. **Modes of Sowing.** The broad cast or hand sowing is the most common for the smaller grains. Another and a better method is by the use of the seed sower or drilling machine. By the first a greater amount of seed is required, while it is difficult, even for a skilful workman, to distribute the seed equally. By the second, the seed

is not only uniformly distributed, but may be sown in drills, which has some decided advantages over broad cast sowing, especially for wheat. Winter wheat sown in the drill is less likely to be thrown out by the frosts, because it is more uniformly covered and better rooted.

501. Any concentrated manure may be put into the ground with the seed, and the growth of the plant may thus be promoted. A larger yield is secured in proportion to the quantity of seed sown, and a larger yield per acre. Drill sowing, or sowing in rows, also allows cultivation by a machine admirably adapted to this purpose, if the crop needs it, during the early part of its growth.

502. When seeds of any kind are sown broad cast by hand, they may be covered by the plough, the harrow, the cultivator or the roller. The larger seeds, like Indian corn, are usually dropped by hand and covered with the hoe, but they may be dropped and covered by seed sowers made expressly for the purpose. When a large extent of land is to be planted the machine is far more economical. Indeed, it is often necessary to use it to save time and labor. Seed sowers are used only on land properly prepared by ploughing, manuring and harrowing. They are made to drop the seed either in hills or in rows, according to the wish of the farmer.

503. If the machine is not used the ground is first prepared by ploughing and harrowing, and furrowed three or four feet apart, according to the kind of corn to be planted, with a light horse-plough; the manure is dropped in the hills at suitable distances, and the seed then dropped upon it by hand and covered with the hoe.

504. It is generally found best, especially on late lands, to spread and plough in a part of the manure, and to drop the remainder in the hills. The manure in the hills

gives the crop a vigorous start at the outset, while that which is ploughed in, being better distributed in the soil, has its effect afterwards, and the crop does far better in the end than it would if the whole were placed in the hill. The land is also left in a better condition for a future crop where the manure is spread. Many use some concentrated manure in the hill, and plough or harrow in the coarser barnyard manures.

505. **Transplanting.** Transplanting is the removal of a plant from the place where it has grown to another. The seeds of many plants, as those of tobacco, cabbages, and many varieties of shrubs and trees, are often sown in a place prepared for the purpose, and the plants springing from them afterwards transplanted to the fields where they are to grow.

506. This mode of culture has several advantages: it confines the expense of the early culture to a small space, while the seed is placed in the best condition for its early and rapid development; it also gives more time for the preparation of the land in which the crops are to be raised.

507. To make transplanting successful, the plants should be strong and vigorous. They may be made so by preparing the seed-bed thoroughly and taking care to prevent them from being crowded by each other or injured by weeds, after they have sprung up. They should be removed very carefully, all injury to the roots being avoided, otherwise they will suffer much from the removal.

508. While the plants are young there is little danger in transplanting, and if they are set out in a mellow and well manured soil at a favorable time, they will continue to grow with only a slight temporary check.

509. In removing older plants, like trees and shrubs, which have been undisturbed for a long time, the utmost care is required in taking them up, to prevent the loss of the small fibrous roots which often extend to great distances from the trunk.

510. The growth of the stem, or that part of the trunk above ground with its leaves and branches, is in proportion to the extent of the roots, and the injury which the latter sustain in transplanting may be counteracted, in a measure by trimming off a corresponding portion of the top.

511. The laceration or breaking of the roots checks the growth of the top in proportion to the injury or loss of the root. In the natural condition of the tree there are only roots enough to absorb the nourishment required by it, and when a part of the root is cut off, or seriously injured, the remaining part cannot, of course, furnish sap enough for the whole tree. In this case, if a part of the top is removed, less sap is required, the remaining roots can supply all that is necessary, and the tree may thus be saved.

512. One method of obtaining good shrubs and trees for ornamental purposes, is to sow the seeds in beds properly prepared. The soil used for this purpose should be deeply trenched and richly manured to promote rapid growth. It is most convenient to lay out the beds from three to five feet wide, and to have the rows run across. Early autumn is generally thought to be the best time for sowing, though some prefer mid-summer. The seeds of each species may be sown soon after they have become fully ripe.

CHAPTER XVI.

CULTURE OF THE CEREALS.

513. The plants generally cultivated by farmers may be divided into four classes: 1. The cereals or grain plants, comprising the plants cultivated for their large farinaceous, or mealy seeds; 2. Leguminous vegetables; 3. Forage plants, or plants used principally in the feeding of stock; and 4. Plants used in the industrial arts.

514. **The Cereals.** The term *cereal* is derived from Ceres, the fabled goddess of corn. The cereals embrace all those annual grasses cultivated for the nourishment of man, including Indian corn, wheat, rye, barley, oats, rice and millet. Buckwheat might be added, in a practical classification, though not properly included among the cereals, as its seeds have much the same quality and are used for the same purposes as those of the cereals properly so called.

515. **Indian Corn,** or maize, is one of the most important of the cereals cultivated in this country, both on account of the numerous uses to which it may be put, and the great amount of nourishment it contains. It is an American plant, and was found in cultivation among the Indians on the first discovery of the continent.

516. Light and porous loams a little sandy, are most likely, if well tilled, to yield large crops of Indian corn. But it easily adapts itself to a variety of soils, and will flourish on all if well manured, except the strongest clays.

517. Land should be prepared for Indian corn, in very much the same way as for other crops, and the preparation

must vary according to the crops for which the piece has been used and the state it is left in. If the field that is to be planted with corn has been in grass for some years, it should be well ploughed the autumn before the planting, and then left till spring, when it will be partially mellowed and may be cross ploughed, manured, harrowed and planted.

518. But stiff, undrained soils, and lands lying on the slopes of hills liable to be washed down by the rains, should, if possible, be broken up in the spring instead of the fall, as the sward will not rot, and if turned up in cross ploughing in spring, will be troublesome during the cultivation of the crop.

519. The manures used with this crop must be varied according to the character of the soil. On light, well worked and mellowed land, old and well decomposed barnyard manure or compost is best, but if the soil is stiffer and somewhat clayey the coarser barnyard manures may be used to advantage, as they improve the texture of the soil and produce heat by fermentation.

520. It is generally thought best to plough in the coarse manures in the fall, as they thus become decomposed and prepare the ground for spring planting. They may be turned under on the sod or on a grain stubble. But if the ground is level they may be spread upon the furrow, after fall ploughing, and left over winter to be turned under in cross ploughing in spring.

521. In cross ploughing, the sod turned under the autumn before should not be disturbed. If the manure is ploughed under in the fall, some finer compost should also be used in spring to be spread on the furrows after cross ploughing, and harrowed or cultivated in. If the soil be stiff and cold, such as is ill adapted to Indian corn,

a portion of the fine manure or compost should be placed in the hill.

522. The Indian corn plant is a gross feeder, and needs a great deal of manure unless the land is very rich. If all this manure is put into the hills, the labor and expense of application and the care of the crop through its whole growth will be increased, on account of the hilling up around the corn made necessary by putting so much in the hill.

523. Another objection to putting much coarse manure in the hill is that the plant is more liable to suffer from drought, and the land is not benefited to so great an extent as when a part of the manure is spread or evenly distributed through the soil.

524. Some spread and plough in the coarser manures, and use some concentrated fertilizer in the hill to give the crop an early and vigorous start. No doubt a judicious use of concentrated manures is good economy, and in some circumstances it may be well to use them more freely, but they are not to be recommended in all cases, as their cost is frequently greater than the profit which may be made from their use.

525. To raise corn profitably the land must be in good condition; it may be made so by the use of a sufficient quantity of manure at the time of planting, or by long-continued and judicious manuring previously. It is not worth while to raise poor crops. It requires about as much labor in ploughing, hoeing and harvesting, to raise thirty or forty bushels per acre, as to raise from sixty to seventy-five bushels per acre, and the profit is greater with the larger crop.

526. In the culture of Indian corn, as of many other crops, the one thing especially important is thorough and

careful ploughing in the first place. There can be no successful cultivation of this crop without it.

527. The land having been fully prepared by repeated ploughing, manuring and harrowing, the next step will be to plant the seed. This may be done by hand or by a machine. If the grains are to be dropped and covered by hand, the rows are marked out by furrows made with a light one-horse plough or some similar implement.

528. The hills should be three or four feet apart in each direction, the distance between them varying according to the kind of corn which is to be planted; the smaller varieties require less space than the larger. If the corn planter, (Fig. 24,) is used, the labor of furrowing is avoided, but with most machines it is difficult to make the rows straight and set the hills at equal distances apart in each direction, so as to be able to run the cultivator or horse-hoe both ways, an important means of saving expensive manual labor.

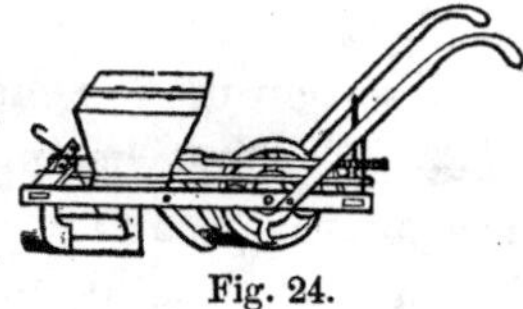
Fig. 24.

529. If the soil has been properly prepared and is in good condition, it is best to have the plants stand as closely as they can without interfering with their perfect development, for it is better that the soil should be well shaded. The spaces between the hills should, therefore, be only just enough to allow the necessary cultivation and the free access of air, light and heat. On poor lands only a smaller number of plants should be suffered to grow, but it is better to put fewer in each hill than to increase the distance between the hills.

530. Many farmers soak the seed some hours before planting, as a means of preventing the depredations of insects, squirrels, or birds. There may be cases where it is

necessary, but except in particular cases this seems to be altogether unnecessary. It may sometimes be useful, however, by keeping off these various depredators. In such cases soak the seed in tar water twelve hours, then coat it with ground plaster, or ashes or lime.

531. Larger crops can generally be obtained by drill planting instead of planting in hills, but the labor of hoeing and cultivating is greater, and except for the smallest varieties, drill planting is not common.

532. Indian corn, whether planted by hand or with a corn planter, should generally be covered about an inch and a half deep to insure sufficient moisture, and give the plant a firm hold on the soil. But on a moist or heavy soil an inch is enough.

533. The first hoeing or dressing may be given when the plants are about two inches high. At this time a light plough may be used, running as near one of the rows as it can without injuring the plants, and then returning between the same rows and running near the other. A back furrow will thus be left half-way between the rows which should not at this time be disturbed by the hoe. The plough will do no injury while the plants are still so small and before the fibrous roots have extended.

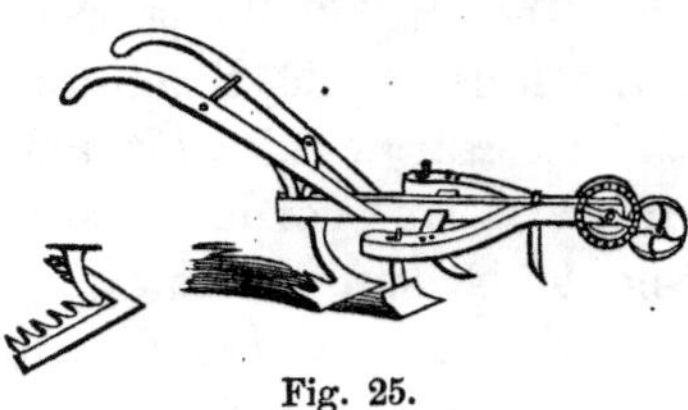
Fig. 25.

534. In subsequent dressings, the horse-hoe, (Fig. 25,) should be used. The plough would break and injure the roots, and should never be introduced between the rows after the first hoeing. The horse-hoe will stir the ground as deeply as it is safe to go. In the second dressing, the cultivator, or what is far better, the horse-hoe, will partially level the back

furrow made by the plough, and a third dressing will leave it quite level.

535. Three hoeings are thought by some to be requisite for Indian corn; but, in general, the oftener it is hoed the better. Should a drought occur, the frequent use of the horse-hoe is particularly advantageous, especially if there be a moist subsoil. It gives the soil a useful stirring and will produce a much more vigorous growth. Great care should be taken that no weeds be allowed in the field.

536. While the crop is still standing in the field, just before the gathering, the farmer should mark the earliest and best formed ears, so that they may be distinguished at harvesting and saved for seed the next year. This is better than to trust to a selection at the time of husking, or after the corn is put into the bin.

537. Those who make a practice of cutting the top stalks, do it about the middle of September, or when the tassel begins to grow dry, after the kernel has hardened. In some cases it is thought that cutting the stalks hastens the ripening of the grain, but if the ears are soft at the time of cutting, they will shrivel and never ripen full and sound.

538. But the best and most enlightened practice appears to be to cut up the whole plant from the ground after the stalk has slightly turned and begun to ripen, and stook it or set it in a cluster of bundles bound together at the top so as to shed the rain, where it will soon ripen up, when the ears may be taken off as it stands on the field, or the whole removed to the barn to be husked.

539. By far the quickest and cheapest way to cut and stook, is to take a pole twelve feet long and fix to one end two legs or supports four or five feet long. The pole is

pierced with a hole through which to insert a cross stick horizontally. Two men take five rows, setting the stooking pole on the middle row, and cut up enough for a good sized bundle for each of the four corners made by the cross stick as shown in figure 26. The binding and a twist around the top of the four bundles, is the work of a moment, when the cross stick is pulled out and the pole drawn along for another stook.

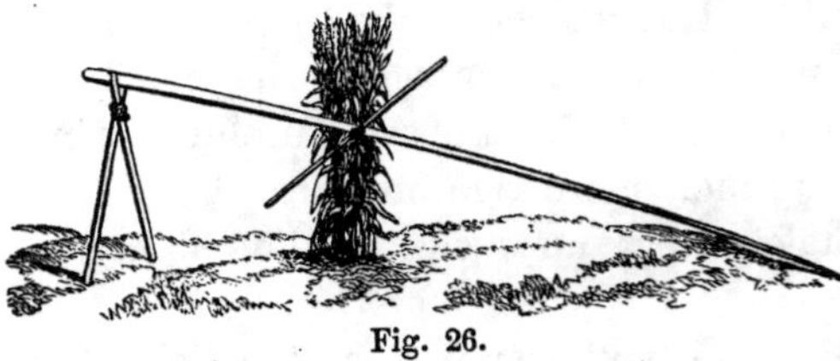
Fig. 26.

540. **Wheat.** There are many varieties of wheat, the differences between them being generally the result of differences of climate, soil and culture; but those most commonly raised may be distinguished by the general terms of winter and spring wheat. The form of the ear is shown in figure 27.

541. The root of winter wheat is most admirably fitted to endure the severe colds of a high latitude. The main seminal root is pushed out at the same time with the germ, and nourishes the plant in its early growth. Winter wheat has a larger and plumper ear and a harder and more erect stem than spring wheat. It should be sown early in autumn, in our latitude as early as September.

Fig. 27.

542. Wheat requires a stronger and more tenacious soil than Indian corn, and more moisture; but if water is found in excess, the tissues of the plant become soft and watery, and it runs to stalk, producing little grain. Soils of a moderate degree of stiffness are best suited to it, but

it will grow on a light soil far better in a damp climate than in a dry one.

543. The soil must of course be such as to furnish the plant with the mineral substances it requires. Lime, for example, in small quantities, is essential to good wheat land, and no soil, however good it may be in other respects, and however favorable the climate, will produce first rate crops of wheat, unless it contain a proper proportion of lime.

544. Though wheat, like most other plants, thrives best on a thoroughly tilled soil, deep ploughing is less important in its cultivation than in that of Indian corn, since its roots do not strike down so deep, while from the season of its growth it is not so liable to suffer from droughts. But thorough cultivation is requisite that the land may be as clean as possible, that is, perfectly free from weeds and noxious plants at the time of sowing.

545. The land having been well manured, ploughed and harrowed, wheat may be sown broadcast by hand or by a broadcast sowing machine, (Fig. 28,) and harrowed in, or it may be sown in drills by a machine admirably adapted to this purpose.

Fig. 28.

546. Both methods have their advantages, but the drill sowing is the more economical of the two, as it saves seed by its more uniform distribution. Wheat properly drilled in is less liable to be thrown out by the frost and killed. The yield per acre is also larger, particularly if care be taken to stir the ground and keep out weeds between the drills during the growth of the plant.

547. Besides, in drill sowing the crop gets the benefit of greater light and heat, and a freer circulation of air, and hence a more thrifty growth. In addition to these important advantages of the drill over hand sowing, some concentrated manure may be applied in the drill, and the wheat feels its influence more directly and quickly than if all the manure were spread and turned under.

548. From four to six pecks of winter wheat, and from two to two and a half bushels of spring wheat, should be sown to the acre. The quantity will vary according to the fertility of the soil, the smaller quantity being required on the most fertile soil.

549. The culture of roots forms an excellent preparation for wheat, because they cleanse and mellow the soil. Wheat should therefore follow a root crop in the rotation rather than an Indian corn crop, though on an oat stubble it is often found to succeed well.

550. Unless the ground has been heavily manured for a previous crop, it should be well manured before sowing wheat. A strong and vigorous growth in the fall is very important, as it will enable the roots to store up a large amount of nourishment for the early spring growth, and the plant will advance with great rapidity in the early part of the following season. Spring wheat should be sown as early in April as the condition of the land will allow.

551. Wheat should be harvested before it gets dead ripe. It makes more and better flour if cut just after the grain has begun to harden, but while it is still so soft that it can be crushed with ease between the thumb and finger. The straw is then greenish but partially turned yellow.

552. If the wheat is not gathered at this time it changes very rapidly in favorable weather, and the grain and straw soon grow less valuable, a part of the starch of which the

grain is composed becoming bran. This should not be forgotten, and when the wheat reaches the proper degree of ripeness it should be cut at once.

553. Exposure to rains after cutting is very injurious to wheat. It makes both grain and straw darker in color and is apt to cause a partial decay on the surface. The parts thus affected mix with the rest in grinding, and give the flour a dark hue. Wheat should therefore be stacked, or housed as soon as possible after reaping.

554. **Rye.** Rye holds the next rank among the cereals in its nutritive qualities and its importance as food for man. The form of an ear of rye is shown in figure 29. It occupies the same place in the rotation on light soils that wheat does on heavy ones.

555. Wheat, as we have seen, is most productive only on a calcareous soil—that is, a soil which contains more or less lime. Rye accommodates itself to much lighter and drier soils, and though it does better where there is some lime in the soil, it does not require the presence of this substance as wheat does, and in point of fact it is usually sown upon the poorest soils of the farm.

556. There are two well-marked varieties of rye, the winter and spring, which are cultivated like winter and spring wheat. Rye is much less sensitive to the cold than wheat, while its growth is much more rapid. Hence it is a better staple crop for a high northern latitude.

Fig. 29.

557. When sown for its grain, about one bushel of seed per acre is required. If sown as a green crop for soiling or feeding out green to cattle, two or three bushels per acre are usually allowed.

558. On sheep farms winter rye sown the previous fall, will often furnish a very early and nutritious feed in spring before the pastures are in a condition to graze, and the more extended use of this crop for this purpose would be judicious, particularly on dry poor soils near the homestead.

559. Rye straw cut short and steamed, is sometimes mixed with Indian or linseed meal, shorts, or other fine feed, and contains more nutriment than the straw of wheat, but it is so tough and coarse that it is not relished by cattle unless artificially prepared, while its value for other purposes is such that it is seldom used as food for stock.

Fig. 30.

560. The principal disease of this plant is known under the name of ergot. It is a kind of spur or morbid growth which takes the place of the grain. Ergot is not confined exclusively to rye, but occasionally attacks some of the other grasses, though more common in rye. It is poisonous. Rye is more liable to it in low damp lands, than on dry and light uplands. It is illustrated in figure 30.

561. **Barley.** Barley (Fig. 31,) grows and ripens with astonishing rapidity, and hence may be cultivated in many climates where other cereals cannot. It requires a light fertile soil well cultivated and free from weeds, which are more injurious to it than to any other grain. The manure used should be old and well decomposed.

562. Barley should be made to follow a hoed crop, if possible, and should be sown as soon after the tenth of May as practicable. It may be simply harrowed in on stiff soils, or harrowed and rolled on light ones. After

coming up it is more likely to be hurt by the feeding and trampling of sheep and other stock than either wheat or rye.

563. It should be harvested before it is perfectly ripe, as it is soon injured if allowed to stand too long. If harvested early, the grain is of better quality and less liable to shell off and be wasted.

Fig. 31.

564. **Oats.** Oats (Fig. 32,) do best in a damp climate and a moist soil, with a moderate summer temperature. As we seldom find these conditions united in this country, the crop rarely succeeds so well here as in some other countries.

565. In the best oat districts of Scotland and Ireland, the average weight of a bushel of oats is forty-three or forty-four pounds, while more than a hundred bushels per acre are often gathered. In this country crops of eighty or ninety bushels are regarded as large, the average yield being much less, while the weight per bushel is rarely more than from twenty-eight to thirty-two pounds.

Fig. 32.

566. After thorough ploughing, oats may be sown broadcast either by hand or by some of the admirable broadcast seed sowers, and covered by means of the common harrow and the roller. The latter is especially useful on light lands, as the compression of the soil affected by it hastens the germination of the seed and causes it to spring up uniformly. From two to four bushels of seed per acre should be used, according to

the richness of the soil and the purpose for which the crop is designed.

567. Oats produce an admirable green crop for feeding out to milch cows and other stock, on account of the rapidity and earliness of their growth. When sown for this purpose a larger quantity of seed is required than if the design is to produce a crop of grain. In either case the earlier they are sown in spring the better.

568. The roller is sometimes drawn over the young plants before they have tillered, or sent up side shoots. It then checks the upward growth of the main stalk and multiplies the side shoots, thus increasing the amount of the product.

569. Oats should be cut before the straw has turned completely yellow; if left longer, the amount of nutriment both in the grain and the stalk becomes less, and there is a loss by shelling out in harvesting. They may be cut with the scythe, and in many cases the mowing machine or reaper can be used to advantage. They should be left to dry a day or two before storing in the barn.

570. In this country oats are used almost exclusively for feeding horses and other animals, for which purpose they are very excellent, as they contain a large amount of nourishment. Oat meal is also extensively used by young men during their training or preparation for athletic games and exercises, being admirably adapted to the formation of muscle and strength. It is used for human food to a great extent in Scotland and Ireland. The straw is more valuable for fodder than that of wheat, rye or barley.

571. **Buckwheat.** Buckwheat is not properly a cereal grain, but belongs to an entirely different order of plants

known as *knotweeds*. There are three cultivated species of this genus, the seeds of which when ground, are used as food for man. Of these only one, buckwheat, is raised in this country, one in Italy, and the third in China.

572. As it remains in the ground but a short time, it may be cultivated in high northern latitudes, and it is seldom found in this country except in the region north of Tennessee and North Carolina.

573. This plant succeeds best on light soils, but will do well on almost any soil except a heavy clay. It is frequently sown to plough in green as a manure in preparing for some other crop. For this purpose it is less valuable than clover, or a suitable mixture of plants, but if ploughed in when in blossom, it is beneficial in all soils which contain but little organic or vegetable matter.

574. Before sowing buckwheat the land is usually ploughed once and then lightly harrowed. No other preparation is necessary. The seed is sown in June, and harrowed in. About three pecks per acre is enough, though some farmers sow a bushel, broadcast. Good crops of buckwheat have sometimes been obtained from seed sown after a crop of barley has been taken from the land, and some sow it in August with winter wheat.

575. When ready for harvesting, it may be cut with the scythe or the cradle; the latter is better. It is then raked or gathered into small bundles, which are fastened by twisting the tops, and allowed to stand and dry on the field. If mown with the scythe and left in the swath, it will shell out. It dries slowly, and should be threshed as soon as it is stored, since there is much danger of its heating. The yield of this crop is from twenty to forty bushels per acre.

576. **Millet.** Several plants of different species pass under the name of millet, and are cultivated, to some extent, for their seeds. The common millet is best known in this country. Millet is often sown to cut up green for stock. If raised for winter fodder, it is cut and cured like hay.

577. Millet flourishes best in a dry sandy loam, well and deeply pulverized by the plough and the harrow. If evenly sown, a peck of seed per acre is enough, if it is cultivated for the seed. But when it is designed to be cut to feed out green to cattle, a larger amount of seed should be used.

578. Millet is regarded as an exhausting crop if allowed to ripen, but it will do well on land too light for grass, and deserves to be more extensively cultivated than it now is. It may be sown from the middle of May to July, and harvested as the grasses are for hay, but when cultivated for the seed, it should be allowed to stand till nearly ripe.

CHAPTER XVII.

LEGUMINOUS PLANTS.

579. This class of plants embraces several different genera and many species and varieties due to the action of soil, climate and cultivation. It includes the cultivated varieties of the bean, the pea, the lentil, the lupine, and the vetch; all of which produce seeds composed largely

of a substance known to chemistry as legumine, which is almost the same as caseine or the cheesy matter of milk, and in many respects is like the gluten or nitrogenous compounds of the cereals, although somewhat different. But the proportion of starch and nitrogenous substances contained in the leguminous plants is far greater than that of the albumen and gluten in the cereals.

580. **The Bean.** The most important of the leguminous plants in our agriculture is the bean. There are many varieties of the bean, all derived originally from the same. The kinds most frequently used belong to the genus *Phaseolus*, of which three prominent varieties are commonly cultivated as a field crop. These are the *large white bean*, the small white, and the China bean, with a red or pink eye. As many as thirty or forty sub-varieties of this genus are found in gardens, some of them known as climbing, or *pole beans*, others as *bush beans*.

581. Beans grow well on a variety of soils, from a very light sand to a strong loam; but sandy or gravelly soils are better for them than strong and tenacious clays. On light soils the plant not only ripens earlier, but is cleaner and freer from earth, which frequently adheres to the plant in large quantities, during rains, especially at the period of ripening.

582. The land should be thoroughly ploughed and harrowed so as to be well mellowed. The stable manure applied should be well decomposed or composted, and it may be placed in the hill or drill. The varieties of the white bean are usually grown in hills, while bush and garden beans are more often planted in drills. On dry, sandy or gravelly lands beans do better if planted thick; the rows of the smaller varieties need not be more than two feet apart, only space enough being left between them

to allow cultivation. In drills six beans may be planted to the foot, and the quantity of seed to be used per acre whether sown in hills or drills, will be from one to three bushels, according to the variety.

583. The proper time for planting beans in the latitude of New England, is between the 20th of May and the 10th of June. Generally the best time is about the 1st of June, but it varies a little, according to the nature of the soil and the forwardness of the season.

584. When the plants have formed their first full-sized leaves, generally about the 20th of June, the crop should be hoed for the first time with the hand-hoe, the horse-hoe or the cultivator having previously been used between the rows, if necessary. The best farmers prefer not to stir the ground with the plough if the weeds can be kept down with the hoe.

585. The character of the season makes a great difference in the crop. Too much moisture causes the leaves to grow with great luxuriance, and a very dry season often stints the plant and prevents it from growing vigorously.

586. When the leaves shrivel and the pods turn yellow, the crop should be harvested, by pulling up the plants and stacking them in some convenient place on the ground or on rails. They will soon become dry, and should then be taken to the barn and threshed out. Unless perfectly ripe and dry, they should be spread out and occasionally turned till all moisture has passed off, so that there is no longer any danger of injury from heating.

587. The yield will vary from fifteen to thirty or forty bushels per acre, according to the land and culture, and the variety planted. The stalks are valuable as fodder for sheep and horses.

588. **The Pea.** The gray or field pea is most common as a field crop. Many other varieties of this vegetable are found in the garden and the market, each of which is marked by some peculiarity as to time of ripening, size, &c.

589. The soil best adapted to the pea is a stiff loam, such as might be called clayey. But it will not do well on a heavy clay. In general the pea may be successfully cultivated on all soils which can be deeply tilled and richly manured, except the stiffest clays and light sands.

590. Fine, well-rotted composts or ashes, plaster or lime, should be used for this crop, in preference to coarse barnyard manures.

591. In soils of not more than ordinary stiffness, which have been well cultivated for some preceding crop, a single deep ploughing followed by the harrow is sufficient for pease. They should be sown in drills, from two to four bushels of seed being used per acre, and covered about an inch and a half deep. They may follow any farm crop in the rotation, but should never be raised year after year on the same land. Many sow pease broadcast with oats, and harrow them in, and good crops are often obtained in this way. A thorough rolling with a heavy field roller is useful.

592. When ready for use pease are picked by hand, or if sown broadcast mixed with some other crop, they are cut with the scythe, and then taken to the barn and threshed with the flail. In some places the pea is cultivated to some extent to furnish green feed for stock, and as a green manure crop to be turned under. For these purposes it is sown broadcast or hoed in among corn at the last hoeing.

593. This plant is liable to be attacked by a weevil, the pea bug, magnified in figure 33, which deposits its eggs in the pod just as the pea is swelling. This is done at night or in cloudy weather. As soon as hatched the grub penetrates the young pea and remains there till towards the end of the following winter, when it bores its way out, after having changed into a pupa and cast its skin, leaving a round smooth hole. The germ is left untouched, and pease injured in this way may therefore be used for seed.

Fig. 33.

594. Immersing the seed in hot water before planting will destroy the grub, if it still remain in the pea, but this remedy would generally be too late, as the grub usually leaves towards the close of winter.

595. The insect lives in other plants, so that if destroyed in every pea there would still be enough left to deposit an egg in every pea of the next crop. Hence there is at present no known remedy against the weevil for early sown pease. Those planted late in June are not so liable to be attacked, and pease might perhaps be obtained free from these insects by late planting.

596. But this vegetable must have abundant moisture while in blossom, or its yield will be small, and the droughts and great heat of July are very injurious to it; hence it will often be found that the evils of late sowing are greater than its advantages.

597. **The Lentil** in some countries forms an important article of food. It requires a warm, light soil, but its yield both of straw and seed is small compared with that of the bean or pea, and there would, probably, be no object in introducing it into our agriculture as a field crop.

598. **The Vetch** would doubtless succeed well here as a green food for cows in milk, or for horses. It might be

sown with oats, using two bushels of vetches, of the white flowered variety, to one of oats per acre, on land in good condition.

CHAPTER XVIII.

ESCULENT ROOTS.

599. **The Potato,** one of the most important plants of the farm, may be raised from the seed, and it is in this way that new varieties are obtained, or it will grow from the tuber or enlarged portion of the stem beneath the ground; this contains many eyes or germs, from which spring shoots or stalks, which reproduce the same species or variety.

600. If the tubers are to be planted, which is the common mode of propagating the potato, it is desirable that they should not be allowed to ripen fully. They grow much more vigorously if dug before ripening than if the plants stand till they decay in autumn.

601. There are many varieties of the potato, but the chief practical distinction is known by the terms *early* and *late*. All the varieties without doubt have come from the wild plants native to South America, whence they were first brought into use in Europe.

602. The potato contains a large quantity of starch in combination with water, and a large percentage of potash which is found in the ash, left after burning. The amount of starch is different in the different varieties, some having as much as thirty-two per cent.

603. The quantity of starch is greatest in winter. Germination rapidly decreases it in spring, and hence potatoes are less mealy and palatable. Since the prevalence of the potato rot, the amount of starch in most of the varieties has very much diminished. It is worthy of remark that the wild potato plant contains but little, if any, nutriment.

604. With good management and in a good season, a fair crop of potatoes may be obtained from almost any soil, but they do best on a loose, mellow, virgin soil, or one newly cleared, and the liability to rot is less in such soils than on a heavy retentive one, or on peat land which before the rot first appeared often produced very large crops. A strong, deep, warm loam with a porous subsoil is especially fitted for this crop.

605. Very few plants require so little preparation of the land for cultivation as the potato, and a large yield has been obtained by merely dropping the tubers along the side of the furrow on the turned up sod, and back-furrowing to cover them.

606. Strongly heating manures, such as that from the barnyard while still unfermented, which were formerly much used for potatoes, have been found by experience to increase the liability to disease, and hence should be avoided, if possible, and if used at all they should be ploughed in rather than applied in the hill. Ashes or plaster of Paris may be used in the hill to advantage.

607. The potato may be cut into pieces before planting, each piece containing one or more eyes or germs, and a certain proportion of the body of the potato. The latter furnishes nourishment to the germ in the first stages of its growth. Cutting is often judicious, and always so when the potatoes to be used as seed are to be

purchased. The largest potatoes grow from eyes taken from that part of the tuber nearest the stalk.

608. The crop may require two careful hoeings, and the weeds should be kept down by further cultivation, if necessary. At the first hoeing, when the plants are from one to two inches high, the plough or the cultivator may be used between the rows, as the workman may prefer.

609. The crop is harvested in the month of September or October, according to location and the variety, being lifted out of the ground by the hoe, or, which is far better, the eight-tined fork. Some farmers run a furrow with the common plough through the rows.

610. **The Turnip.** The turnip is cultivated with the highest success only in a moist and equable climate. In this country, on account of the excessive droughts to which we are subject, the large size of root and luxuriant growth so frequently found in Scotland and the west of England, are seldom to be seen. Possibly the deficiency in weight of the crop may be made up by a greater amount of nutriment in proportion to weight, as in the case of grasses and other plants grown in a dryer climate. But this must be determined by more extended experiment and accurate analysis.

611. The common turnip is very highly esteemed as a valuable food for stock, especially for sheep, and its cultivation is regarded as one of the best methods of preparing the soil for a succeeding crop of grain.

612. Experience has shown that it is very advantageous to raise alternately a deep or tap-rooted crop like the turnip, carrot or parsnip, and a surface-rooted one like wheat, rye, barley, &c. The form of the root of some of these plants is shown in figure 34. The root crop is

Fig. 34.

not only valuable in itself, but it also draws up from the lower strata of the soil more or less of the valuable plant nourishing substances always present there, and leaves a portion of them near the surface, where they can easily be reached by surface-rooted plants.

613. The varieties of the turnip are very numerous. Those most commonly cultivated are the common globe, the purple-top strap leaf, the hybrid, and the Swede or ruta-baga. Many others have a local reputation, and are more or less valuable.

614. The soils best adapted to the turnip are light loams, loose and open, under full cultivation or thoroughly ploughed and pulverized. There are few crops which require so much preparation of the land before planting.

615. The land designed for the Swede or ruta-baga, should be very deeply ploughed the preceding autumn, the deeper the better. Two thorough ploughings should also be given in the spring, to be followed by a careful harrowing so as to mellow and completely disintegrate or break up and pulverize the soil. The flat turnip requires less depth and thoroughness of cultivation.

616. The soil should be enriched by an abundant supply of manure. On poor soils the root soon degenerates and becomes small and acrid. The manures best adapted to this vegetable are those rich in phosphates, such as dissolved bones or bone dust, guano and super-phosphate of lime.

617. Manures rich in nitrogen and comparatively poor in phosphates, promote the growth of the leaf rather than of the bulb, and their injudicious use will produce an inferior crop. When the soil is not very rich and soft in itself, a heavy dressing of farmyard manure may safely be ploughed in, and home made super-phosphate or bone dust, mixed with guano, may be applied near the surface or in the drill.

618. The common round or flat turnip is usually sown broadcast and harrowed in, but the Swede or ruta-baga is sown in drills about two and a quarter feet apart, with the seed sower. Neither should be planted in ridges or raised drills, except on very thin soils, as the benefit to the land of a deep-rooted crop is less marked, than if the ground is kept level.

619. From two to three pounds of seed are allowed per acre. This quantity will give more plants than can be grown to advantage, and they should be thinned out so that there may be a proper distance between them during the summer.

620. The horse-hoe may be used between the drills when the first rough leaves have appeared. This is followed by the hand-hoe to clear out the weeds and stir the soil around the plants. Subsequent hoeings will be necessary to prevent the growth of weeds.

621. Turnips may remain in the ground till the hard frosts begin, without injury. They should then be taken up and stored in suitable root cellars or in pits on the field, where they may remain till wanted for use.

622. As has been said, turnips are a valuable article of food for sheep and all kinds of store cattle. An animal can easily be fattened on turnips and hay. They should be cut with the shovel or the turnip-slicer before being

fed out. From seventy-five to one hundred pounds a day, in addition to hay or straw, may be fed to an animal of a thousand pounds weight.

623. The kohl-rabi is a hybrid turnip, or turnip-stemmed cabbage, much used in some countries as food for man and animals. It is sown early in spring and cultivated like the cabbage.

624. The cabbage is not very common as a field crop in this country, but is mostly confined to the home or market garden. It requires a very rich clayey soil and high cultivation. The seed is usually sown in beds to be transplanted into hills, where it is hoed and cultivated like other garden vegetables.

625. **The Beet.** There are many varieties of the beet, but all may be included under the two general designations of garden and field beets; these may be again sub-divided according to their size and color, the shape of the root, and the purposes to which they are applied. Field beets comprise those used for feeding cattle and making sugar.

626. The **Mangold Wurzel** is more esteemed for stock feeding in this country than any other variety of beet. It does best on a rich, deep, well-manured soil, with thorough cultivation, but will accommodate itself to most soils that are strong, deep, and well tilled.

627. To prepare the land for the beet it should be deeply ploughed, manured, and harrowed level; the seed should then be sown by a machine in rows at the rate of three or four pounds per acre, and covered to the depth of an inch. It is a common practice to steep the seed in water for twenty-four hours before sowing.

628. The after cultivation consists mainly in the free use of the cultivator or horse-hoe, and the hand-hoe, so as to keep the surface fresh and free from weeds. Man-

golds may stand a foot apart in the rows. If they are a foot apart in the rows, the rows being two feet apart, there will be more than twenty thousand plants to the acre.

629. The Mangold may be harvested in October. If the root is bruised or injured it is liable to decay, and care should be taken to guard against the possibility of this. When well stored in a cool cellar or in pits dug for the purpose, it will keep through the winter, and cattle of all kinds are very fond of it.

630. **The Carrot.** The carrot is very valuable as a forage crop, and is extensively cultivated and highly esteemed. No root is more relished by domestic animals. Weight for weight it is somewhat less nutritive than the potato; but its greater yield per acre more than makes up for the difference in quality.

631. Horses are especially fond of it, and when not kept at very hard work, should have it as part of their regular food. It keeps up their condition, and gives them a fine glossy coat. When fed to cows it increases the richness of the milk somewhat, and is supposed by some to give a richer color to the butter, while for sheep and lambs it is also a valuable article of food.

632. The cultivation of the carrot is generally more expensive than that of most other root crops. It requires much slow and toilsome hand labor, unless great care be taken to avoid sowing the seeds of weeds with the manure. But on clean land, and with the use of concentrated manures like ashes, plaster, guano or old and well decomposed compost, the cost of the crop need not be much greater than that of other roots.

633. There are several varieties of this root, all of which probably came from the common wild carrot of Europe, the *Daucus carota.* The most valuable for field

culture are the short horn, the long orange, the white Belgian, and the altringham. The white Belgian will give the heaviest yield, on the whole, but the long orange sells better and is somewhat more nutritious. The white Belgian is often of greater size, but coarser and of less weight in proportion to its size. But many think the short horn yields a more valuable crop than either.

634. The carrot grows in almost any variety of climate found in this country, but it is more especially adapted to the northern regions, which ordinarily suffer less from drought. Excessive dryness stops its growth and materially lessens its product.

635. It is most productive on a deep, light, warm loam, capable of retaining a moderate degree of moisture in summer, but with a dry and open subsoil.

636. Deep ploughing and subsoiling are especially important in the cultivation of this crop. The size and weight of the root depend very much upon deep tillage.

637. No manures of a coarse or very stimulating nature should be used. They cause a useless growth of fibrous roots and leaves to the injury of the main root. Land enriched by previous high culture, where manure will be unnecessary, is to be preferred for this crop, but in any case only old and well-rotted manures, or some concentrated fertilizer, should be used. These may be spread on the furrow after deep and thorough ploughing, and harrowed in when the land is ready for the seed.

638. The seed should be new and fresh. When two years old it will often fail to germinate. As it does not start till after it has been exposed to moisture for some time, it is often soaked for eight hours or more, and then spread out quite thickly on the floor, where it is left till it begins to germinate. This will generally be in six or

eight days. It should then be immediately rolled in plaster and sown by the seed sower, in drills from fourteen to eighteen inches apart.

639. If the seed is new and good, two or three pounds to the acre are quite enough to plant. If its quality is unknown, four or five pounds may be used and the plant thinned out while growing. The covering should be but slight, not more than half an inch in depth.

640. The ground should be fully prepared in the previous autumn, and the seed put in as soon after the 15th of April as possible. The plant does better if started while the ground is still quite moist, since it is very slow in its early growth.

641. When the plants are well up so as to be distinctly seen, they should be hoed and weeded. It is much easier to keep the weeds down at the outset, than to get them out after they have overrun the crop. The number of hoeings will depend much upon the character of the soil and the previous culture. If the land is foul or very weedy, it will require constant and repeated labor, at an expense greater in some cases than the value of the crop itself.

642. At the second hoeing, or when the plants are two or three inches high, they may be thinned out if they require it, but a greater weight per acre may be obtained without much thinning, and the smaller roots, though they do not look quite so well, and will not sell for so high a price, perhaps, are better for stock than very large ones grown four or six inches apart.

643. Carrots may be allowed to stand till the early part of November without injury from frost. They may be raised from the earth by the plough or the fork, and stored for winter use, the tops being fed to stock.

644. **The Parsnip.** The parsnip is another plant which has been made valuable by culture, the original wild parsnip being altogether worthless. It is cultivated both as a field and a garden crop, and deserves far more attention than it now receives from the farmer.

645. There is little doubt that the parsnip is more nutritive than the carrot, that it is more hardy, somewhat less liable to be injured by diseases or insects, while it is more easily cultivated and more productive. It is much liked by all animals, and is thought to give a richness to the milk of cows which no other root can, except, perhaps, the carrot. It is claimed that its use enables the farmers of the islands of Jersey and Guernsey to make butter in winter, as rich and high-flavored as they can upon the grasses of June.

646. There are two varieties of this plant, both derived from the same source. They are the round or garden, and the long field or large Jersey parsnip. The farmer will find the latter the most profitable.

647. The parsnip prefers a mild and moist climate for its early growth, but it endures our severest cold, and may remain in the ground through the winter to be dug up fresh in the spring and used for feeding stock.

648. It is most productive on chalky or clayey soils, and sands rich in mould or humus, but will grow well wherever carrots will. In some parts of France carrots and parsnips are cultivated together.

649. The parsnip being a tap-rooted plant, the soil must be prepared for it in the same manner as for carrots. The seed used should be of the growth of the preceding year. The sowing and after cultivation are like those of the carrot.

650. In a proper climate and soil, the parsnip yields more than the carrot, but it is, probably, a more exhausting crop.

651. The **Jerusalem Artichoke.** The Jerusalem artichoke is nearly as nutritious as the potato, and its stalks are almost as valuable as its tubers. It has never been cultivated to any great extent as a field crop, in this country, but many cultivators of it in Europe claim that it has many advantages. Among others, that it grows well on light sands and tenacious clays, where no other root crop would succeed. They say it does not exhaust the soil, but may be grown year after year in the same place; that it is free from diseases, and endures alike the colds of winter and the droughts of summer.

652. Its cultivation is much like that of the potato, the land being prepared and manured in the same way. The tubers are planted early in spring, in rows or drills, the rows being far enough apart to allow working between them, and the plants about nine inches apart in the rows.

653. In countries where this plant is cultivated as a field crop, the stalks are either cut and fed out green, beginning, in France, about the end of August, or left to be cut with the sickle, and stooked and dried for winter fodder. After the stalks are cut and removed, the tubers are taken up as they are wanted to feed out, or dug late in the fall and stored for winter use. Most kinds of farm stock are very fond both of the stalks and the roots.

CHAPTER XIX.

THE GRASSES — FORMATION OF MEADOWS OR UPLAND MOWINGS.

654. The culture of the natural and artificial grasses and other forage plants arose from the necessity of providing sustenance through the winter, or inclement season, for the domestic animals on which the success of agriculture so much depends. It is evident that this department of farming is of the highest importance, especially when we consider how dependent the raising of stock must be upon it.

655. The grasses may be classed, for convenience, under two general divisions, the natural and the artificial. The natural grasses comprise all the true grasses, or plants with long, simple, narrow leaves, and a long sheath divided to the base, which seems to clasp the stem, or through which the stem seems to pass. Each leaf has many fine veins, or lines running parallel with a central prominent vein or midrib. The stem is hollow, with very few exceptions, and closed at the joints.

656. The artificial grasses are mostly leguminous plants, with a few others which are cultivated and used like the grasses, though they do not properly belong to that family. The clovers, lucerne, sainfoin, medic and other similar plants, are included among the artificial grasses.

657. Lands laid down with the natural grasses are designed as more permanent mowings than those sown with the artificial ones alone. They are sown with a number of species of the true grasses, most of which are

perennial, and are to be used as mowing lands or for pasturage. The artificial grasses are more frequently intended to occupy the ground for one or two years only in the rotation with other crops, and are generally composed of only one or two species of plants, and those annuals, or at most biennials.

658. In this country it is common to sow one or more species of clover with the natural grasses. The clover then occupies the ground almost exclusively during the first and sometimes the second year, but afterwards the perennial grasses take its place and form a permanent turf.

659. The natural grasses form a close turf or sward, and when left uncut to be fed off by animals, this turf makes what is called a pasture or pasturage.

660. There are certain situations which must be improved as pasturage, if at all. Such are steep slopes on which cultivation is difficult or expensive, and where the soil would be washed into the valleys below, if broken up by the spade or plough; also lands which lie along the margins of streams or rivers liable to periodical overflows, by which growing crops might be endangered or the soil be washed away, and low marshy lands which cannot be drained so as to produce annual crops. In these latter situations, however, the wild grasses frequently come in so luxuriantly, on account of the richness of the soil, as to give good crops for hay for many years in succession, without any cultivation whatever.

661. There are great differences between the different species of grasses. Some are short lived, others more permanent; some mature early, others later; some contain much nutriment, others little. The different species require different kinds of soil also, and withdraw from it different substances and elements.

662. By the use of many judiciously selected species together, a greater weight of grass and hay can be obtained from an acre than if only a few species be used. Probably this arises from the fact stated above, that the different species use different kinds of nutriment. On a certain space, say on a square foot of soil, as many plants of a particular species of grass will grow as can find there the kind of nourishment they need; no more of that species can grow there, of course; they would starve as it were, but other plants of a different species of grass, which require different substances to support them, may grow on the same soil, because the plants of the first have not consumed any of the substances which they want; so as many plants of the second species will grow there as can obtain the sort of nourishment suited to them; a third species, and others needing different kinds of nutriment may be added, and this may go on till the soil is crowded as thick with the plants as they can grow.

663. In selecting a mixture for mowing or for pasturage, regard should be had to the modes of growth and other peculiarities of each kind. A grass well adapted to cut for hay, may be very unsuitable to form a pasture turf. Timothy, though one of the best of our grasses for mowing, is not good to sow for pasturage, as it cannot bear the close cropping of cattle.

664. Among the grasses which may most profitably be cultivated for mowing, may be mentioned Timothy, redtop, white bent, orchard grass, perennial rye grass, June grass, rough stalked meadow grass, fowl meadow grass, meadow fescue, and tall fescue.* Other

* The natural history, culture and economic value of the grasses are fully stated in the Treatise on Grasses and Forage Plants, which those who desire to make themselves more familiar with the subject may consult.

species might be mentioned as worthy of cultivation for this purpose in particular localities, or when the hay is to be applied to some particular use, but the above are the most valuable.

665. Among the species more particularly fitted to form pasturage, are meadow foxtail, orchard grass, sweet scented vernal, June grass, redtop, meadow fescue, and yellow oat grass.

666. In selecting the species to be sown, the time of flowering of each species should be regarded. When seeds of different grasses are mixed for mowing land, such kinds should be chosen that all will come into flower at about the same time, otherwise one species will have begun to spoil before another is ready for cutting.

667. In laying down pasture land on the contrary, the object is quite different. Here we wish a constant succession of green and succulent herbage from early spring to late autumn. Hence some species may be valuable not for their nutritive qualities, but from their habit of very early or late growth. The sweet scented vernal, one of our earliest grasses, is an instance of this.

668. The grasses attain their utmost luxuriance only in a moist and mild climate. Severe heats and long protracted droughts check their growth and make it very difficult to form a close sward. Generally speaking, our grasses suffer much more from the droughts of summer than the colds of winter. It should be added that grasses grown in a dry climate, or a dry season, contain more nutriment in proportion to their weight.

669. The best time for sowing the natural grasses, in the latitude of the northern States, is about the first of September, since they can then become strongly rooted before the approach of winter. The practice of sowing

in spring with oats or some other grain formerly prevailed, but the droughts of summer very often killed out the young plants, made tender and weak by the shade of the grain crop, and great losses were the consequence.

670. To form a good seed bed it is desirable that the land should be under cultivation and well manured for two or more hoed crops. It is then deeply and thoroughly ploughed and harrowed, so as to leave it in a mellow and friable condition.

671. The seeds mixed as already recommended, may then be sown by hand and simply rolled in. They should not be covered to any considerable depth, and a heavy harrow will bury many of them too deep. If no roller is at hand, or if the ground is so wet that it cannot be used to advantage, its place may be supplied by a bush harrow.

672. It has been found by experience that in general the grasses do better when sown in the fall by themselves; but on clayey, undrained soils, where fall sowing is impracticable on account of the great liability to injury by being thrown out by the frost, it would be better to sow with wheat or barley in the spring. Such lands will not be liable to suffer from drought.

673. If clover is to be sown on land laid down to grass in September, the March following is the best time. The seed may be strewn on the last light snows of that month, and will vegetate without any covering, though if the land be sufficiently dry a roller may be passed over the surface and will be beneficial.

674. The artificial grasses comprise red, white and other clovers, lucerne, sainfoin, medic and some others.

They may be grown alone or mixed with the natural grasses.

675. Red clover is one of the most valuable and economical of forage plants. Its long tap-roots loosen the soil and let in the air, while by their chemical action they fix gases which enrich the earth very much. The decay of them in the ground also fertilizes it, and the plant shades and protects the surface, and helps to destroy many annual weeds.

676. Clover is what may be called a lime plant, and the soils best adapted to it are clayey or tenacious loams. It generally does well on good wheat lands. Recent investigations have shown that lime enters largely into its composition.

677. White or Dutch clover is as common as the red, and often forms a considerable portion of the turf of pastures of a moist and tenacious soil. It is most commonly cultivated for pasturage, and many think it to be as valuable for that purpose as red clover is for hay, or for soiling or feeding out green to stock; but cattle are not so fond of it.

678. Neither lucerne nor sainfoin are cultivated in this country. The former has been found to be ill-adapted to our climate, suffering severely in the southern States from long continued droughts, and as severely in the northern from the low temperature and the sudden changes of winter.

CHAPTER XX.

PLANTS USED IN THE ARTS AND MANUFACTURES.

679. Plants used in the arts are most commonly divided into three classes: 1. Oleaginous plants, or those raised especially for their oils; 2. Textile plants, or those raised chiefly for their fibre; and 3. Plants used in the processes of dyeing, tanning, and various manufactures.

680. The only plant raised to any extent in this country for its oil is flax, which is also cultivated for its fibre. The seed is ground and the oil pressed out, leaving what is called linseed cake, which when ground or broken up fine is known as linseed meal, a valuable food for stock. The oil obtained from it is known as linseed oil, extensively used in mixing paints and for other purposes, and always sells readily at a good price.

681. Flax flourishes in a great variety of climates, and as it grows very rapidly and requires but a short time to complete its growth, may be cultivated in high northern latitudes. The soil on which it is sown should be rather light, or at least not very stiff and heavy. A light loam inclining to sand, which may be deeply and easily tilled and kept clean of weeds, is best.

682. But the choice of soil should depend on the object in view. If flax is raised principally for the seed, it can hardly be too rich and well-manured. But if the plant be grown mainly for fibre, a very rich soil is objected to, as it makes the fibre rank and coarse.

683. Old and well-rotted barn manures may be used for this crop, and lime, ashes, or other substances

abounding in lime, are good. A heavy dressing of stable manures may also be ploughed in deeply in the fall. In the cultivation of flax it is very important that the lower strata of the soil should be in good condition.

684. If the soil be mellow and under good cultivation, one ploughing followed by a thorough harrowing will be sufficient, but if it be stiff and ill prepared, two ploughings at least will be necessary.

685. The quantity of seed to be sown also depends upon the object in view. If it be desired to raise the seed, only two bushels per acre will be enough. If the fibre, about three bushels is needed. If the less quantity be used, the plant will grow stalky and branch and produce much more seed.

686. But if the larger quantity be sown, the plants force themselves up in a single stem, without branches. This gives a better fibre, as branching shortens it and makes it irregular. A long, straight, fine and delicate fibre is by far the best, and it is found to be more profitable to cultivate the plant so as to obtain this, than to raise it for the seed.

687. The seed is sown broadcast and covered with a light harrow, then rolled. After the plants are up they should be kept as free as possible from weeds, which should be pulled up by hand. If the flax has been sown thick on land well-cleaned by a hoed crop the previous year, the weeds will not be troublesome unless their seeds have been sown in the manure.

688. The old method of harvesting flax was to pull it by hand, tie in small bundles, and stook it. But the processes of manufacture are now so far perfected that the crop may be cut with the scythe or the cradle. The old processes of water rotting, breaking, swingling, &c.,

are now superseded. For the fibre the plant is cut as soon as the blossoms begin to fall, but if the object be to secure both seed and fibre, it should be left till the bolls have turned yellow.

689. When the flax plant is cultivated for the fibre, from ten to fifteen bushels of seed may also be expected per acre, depending on the character of the land and the thoroughness of culture.

690. Hemp, another textile plant, is cultivated principally for the sake of its fibre, which is used in the manufacture of ropes and coarse cloths. It belongs to the same family of plants as the hop and the nettle.

691. The soil best adapted to hemp is a deep rich mould of loam and vegetable matter, with fine sand and clay intermixed. The rich alluvial lands of Kentucky, Missouri, and other western States, are admirably fitted for it.

692. The seed is sown broadcast early in spring, at the rate of from one and a half to two and a half bushels per acre, according to the fineness of the fibre desired. Thick sowing, as in the case of flax, produces a finer fibre. When the blossoms begin to fall in July or August, it is cut up and sorted into different lengths, and bound up into bundles six or eight inches in diameter, and put into pools or cisterns of water for rotting. After being sufficiently rotted, the bundles are taken out, dried and stacked, till ready for the mechanical processes of breaking and manufacture which follow.

693. **Osier Willows.** Among the plants used in various manufacturing industries, and which form a considerable item in the agricultural interest of the country, may be mentioned the Osier willow, broomcorn, and the hop.

694. Osier willows are cultivated for the purpose of basket making. Among the varieties most approved are those known as the Dutch willow, the purple willow, the round-leaved, and the long-leaved triandrous willow.

695. Willows will grow in a great variety of soils if they be only moist enough; but deep, rich, moist intervals or low alluvial lands, lying on the margin of streams, especially such as have a southern exposure protected from high winds, are most suited to them.

697. The willow grows well on moist soils, but it should not be too wet, and in many cases draining the land is advisable, so that it may be ploughed deeply and prepared as if for corn or any other highly cultivated farm crop. It is then ready to receive the cuttings.

697. The slips or cuttings are about two feet long, and should be set perpendicularly in the soil one foot apart, in rows about three feet apart. They should be kept clean of weeds the first year or two, either with the hoe or the cultivator. The osiers may be cut for the first time in about two years after they are set, and may afterwards be cut annually early in the spring.

698. **Broomcorn** does best in a deep, warm, alluvial soil, such as is best suited to Indian corn. The land should be ploughed in the fall, if sward land, and cultivated in spring, or well harrowed and prepared very much as for Indian corn. The seed is sown with a seed sower as early in spring as practicable, in hills about two and a half or three feet apart. It is hoed and thinned out soon after coming up, six or eight stalks being left in each hill, and afterwards cultivated between the rows once or twice in the season.

699. When the season is sufficiently long, broomcorn is allowed to grow until the seed is ripe and hard. It is then

lopped or tabled about two and a half feet from the ground, and the top or brush end, with about eight inches of the stalk, are cut off and laid on the tables to dry. It is then stored on open scaffolds under cover until a convenient time, when the seed is scraped from the brush by drawing it through two steel springs. The brush is then bound in bundles of about ten pounds weight, and is ready for market. The seed is valuable for feeding stock.

700. **The Hop** has generally been considered a valuable crop, profitable in localities where the soil and exposure favored its growth. The most esteemed varieties are the golden, the yellow grape, and the Farnham.

701. The hop requires a deep and rich loam, rather stiff than light, and containing a large proportion of organic or vegetable matter. A dry porous subsoil is also desirable. The quality of the hop will depend much on the soil. It does best in a moist climate.

702. The land devoted to hops should be richly manured, and the use of large quantities of well-rotted barnyard compost, bones, woollen rags and other rich fertilizers, cause it to produce full crops of the best quality.

703. The roots of this plant extend very deep into the soil, and the land should therefore be very deeply ploughed and completely pulverized. The hop is propagated by cuttings or layers, sometimes by sowing the seed. Cuttings which have been rooted in the form of layers, grow more rapidly than more fresh ones.

704. The bines are supported on poles set into the hills. The poles should be from twenty to twenty-five feet long. It is thought by the best hop growers, to be a great mistake to use poles of only twelve or fifteen feet in length as many do, for in general, the yield is much less, and the quality is not so good, while the labor of

hoeing and picking is as great as with the longer poles. Indeed, it is very seldom that a large crop of first rate hops is obtained from short poles.

705. Hops should be gathered when fully ripe, in August or September. The vines are cut off from one to three feet from the ground, and the poles pulled up and laid over large boxes. The hops are then to be picked perfectly free from leaves and stems, dried in kilns, and pressed into bales.

706. **Tobacco** is sown in beds made very rich by manuring, to be transplanted in June, or when the leaves are two or three inches long. The soil may be prepared by ploughing in old and well-rotted stable manures, guano and other stimulating fertilizers.

Fig. 35.

707. Tobacco should be planted early, that it may be cured while the weather is still warm and dry. It is only in this way that a fine quality can be secured. Constant care is necessary to prevent injury from the tobacco worm, shown in figure 35. For this purpose the plants must be frequently examined and the grubs picked off by hand and destroyed.

708. While still in blossom and before the seed has formed, the plants should be topped, about two and a half feet from the ground, leaving twelve or sixteen leaves to the stalk, and all side shoots broken off.

709. When the leaves are thick and spotted, and crack if pressed between the thumb and finger, they are ready for gathering. The plant is then cut, left in the row till the leaves are wilted, and then carried to sheds to be hung up to dry from five to ten weeks.

CHAPTER XXI.

OF ROTATION OF CROPS.

710. By **Rotation of Crops** is meant raising a series of different crops in regular succession. A farmer turns up a lot of his pasture land, and raises, this year, a crop of potatoes; next year, on the same land, a crop of corn; next, a crop of rye; next, clover and grass. This is a common four-fold rotation.

The **object** of rotation in crops is to make a field or a farm, yield, with a certain amount of labor and of manure, the greatest possible amount of valuable crops, with as little exhaustion of the soil as possible.

The **reason** for a rotation of crops is that no two plants, of different kinds, require the same substances, in the same proportion, for their nourishment. The grains and the grasses may soon exhaust the supply of silica. They should, therefore, not immediately succeed each other in a rotation. They should be each followed by a crop which needs less of silica but more of potash or some other mineral salts. A field which would not yield a second good crop of wheat, may, even without manure, give a very good crop of clover, of turnips or of carrots.

711. The **Important Principles** in the rotation of crops are 1st, that though a soil may contain all the mineral substances necessary for the nourishment of every variety of cultivated plants, there is only a limited supply of the mineral food necessary for a particular plant; 2d, some plants, like the grains, draw their nourishment from near the surface; others, like carrots and parsnips, draw

much of it from a greater depth; 3d, some plants, those, namely, which have abundant foliage, draw much of their food from the atmosphere; others, like the grains, depend more upon the materials contained in the soil. 4th, Particular insects live upon certain kinds of plants, certain flies, for example, on grains and grasses, and continue to multiply as long as the same crop occupies the soil from year to year. But when a crop intervenes on which these insects cannot live, as beans or turnips, after wheat or oats, then they perish for want of proper nourishment for their young.

712. The **order** in which crops succeed each other is often of great importance. Weeds are a great injury to all crops, and barnyard manure almost always carries with it the seeds of many pernicious weeds. Such manure should therefore be put into the ground when a crop is to be cultivated, like corn or beets, which may be kept free from weeds by the hoe and the plough or cultivator. When the weeds have been destroyed or nearly destroyed, by a hoed crop, a crop may follow of grain or clover which cannot conveniently be weeded.

713. Much may be saved by rotation. Each crop, in succession, may find in the soil valuable matters which were unnecessary to the preceding crops. Time may be saved, which is more valuable than any crop, for *lost time is never found again.* We must ascertain what is the best succession of crops, and so arrange the different crops in the different fields, as to occupy all the time of the husbandman and yet not give him too much to do at any one time.

With sufficient forecast, this may always be done. Suppose you can keep under cultivation twenty-eight acres. You divide them into seven equal portions, and, if your rotation is one of five years, with grass for two years, call your several fields A, B, C, D, E, F, G. A natural arrangement may be something like the following:—

Years.	On A.	On B.	On C.	On D.	On E.	On F.	On G.
1st, .	Potatoes.	Corn.	C. T., B.*	Rye.	Clover.	Grass.	Grass.
2d, .	Corn.	C., T., B.	Rye.	Clover.	Grass.	Grass.	Potatoes.
3d, .	C., T., B.	Rye.	Clover.	Grass.	Grass.	Potatoes.	Corn.
4th, .	Rye,	Clover.	Grass.	Grass.	Potatoes.	Corn.	C., T., B.
5th, .	Clover.	Grass.	Grass.	Potatoes.	Corn.	C., T., B.	Rye.
6th, .	Grass.	Grass.	Potatoes.	Corn.	C., T., B.	Rye.	Clover.
7th, .	Grass.	Potatoes.	Corn.	C., T., B.	Rye.	Clover.	Grass.

* Carrots, Turnips, Beets.

In this way you always have eight acres of grass, four of corn, four of rye, four of clover, four of potatoes, and one or more each of beets, of carrots, and of turnips.

714. Or you may have a still longer rotation, introducing, after carrots, parsnips, after beets, cabbages, and after turnips, pease and beans.

You may save time in the management of the manure. This may be put in very abundantly before ploughing, and also in the hills for corn, and before beets, carrots and turnips, or before cabbages, parsnips, and pease and beans; thus being put in once or twice only in the whole course. If mineral as well as other manures are used, with the potatoes may be applied plaster, bones and ashes; with the corn, barn manure; with beans, abundant plaster; with the roots, guano, or sea manure, common salt, plaster, bones dissolved in sulphuric acid, and ashes.

In the cultivation of these crops, the ground between the rows should be turned over and stirred as often as possible,—six or eight times at least. The ploughing will then come in this order: earliest, for carrots and beets, next, for potatoes, next, for corn, next, for turnips. The grass-field, for potatoes, may be turned over after hay-harvest; the ploughing for parsnips may be later and the seed be sown in the autumn. All the hoed crops should be cultivated with the cultivator, the horse hoe or the horse plough, whenever time can be found till the crops are too far advanced to admit of it.

Good reasons can be given for the seven-years' course here recommended. Potatoes require large portions of the alkaline salts and of lime. These are succeeded by corn, which requires more of silica, together with alkalies; then come the roots, which require lime and the alkalies, with a good deal of nitrogen; after them rye, which calls for silica. This is succeeded by clover, which demands a great deal of lime, and this, by the grasses, which again demand silica and the alkalies.

715. All plants require, but in different proportions, carbonic acid, phosphoric acid, sulphuric acid, the alkalies, potash, soda and ammonia, and lime, magnesia and iron. The acids combine with the other elements of fertility, and, while the corn is growing, they are preparing, from the particles in the soil, carbonates, phosphates and sulphates of potash, soda, ammonia, lime, magnesia and iron, for the other crops, and new supplies of silica for the grasses.

The substances most frequently needed for the restoration of fertility are ammonia, phosphoric acid and potash, and the most valuable manures, next to barn manure, are accordingly bones and ashes.

18 *

716. On a field of clover, gypsum in powder, ashes, and bones dissolved in sulphuric acid with one hundred times its quantity of water, always produce gratifying effects. On grass land similar effects are produced by the use of liquid manure which has run from the manure heap.

A poor gentleman in Maryland, suspecting that land which had been worn out by long continued cultivation of tobacco, might be restored by plaster, so as to produce wheat, tried the experiment, which was completely successful. He bought many acres of exhausted tobacco land, and, by fertilizing it with plaster, made himself a rich man.

This gentleman, not a man of science, was led to make the experiment by reading Sir Humphrey Davy's Chemical Lectures.

717. **A Fallow.** A field is said to be fallow for a year, when no valuable crop is raised upon it. Such a year is called a year of fallow, and the field itself is sometimes called a fallow.

It is sometimes well to let a field lie fallow. A field much infested with weeds may be allowed to lie till the weeds are well grown or beginning to blossom, when they may be turned under with a plough. This is like giving a coat of manure. When another crop of weeds has sprung up, they may be ploughed in, and this may be repeated as often as there are any weeds to turn under. These green crops may be advantageously increased by harrowing in, after each ploughing except the last of the season, seed of some rapidly growing plant, like buckwheat. After a year of such fallow, the field will be likely to be comparatively free from weeds, as most of the seeds of weeds will have sprouted and been destroyed.

718. Another benefit comes from the fallow, **Weathering.** The soil, often turned up, is exposed to the influences of the air, and to sunshine, rain, cold, and wind. From the air it receives oxygen, carbonic acid and ammonia, which are either employed in rendering soluble the mineral salts lying in the soil, or are laid up in the geine of the soil for the use of future crops.

These salts lie concealed in small stones or minute particles of the rocks. In mica and felspar, for example, which are ingredients in granite, there are potash, alumina, magnesia and iron, as well as silica, and sometimes soda and lime, all essential elements in the food of plants.

719. The old Greeks and Romans often allowed their fields to lie fallow, and found them thereby rendered more fertile; and the same is done, for the same reason, by many nations in the South of Europe. But the introduction of Indian corn, potatoes and other roots, has rendered it less necessary, and where land is very valuable, fallows are generally discontinued, the benefits of weathering being secured by deep ploughing and by frequent tillage between the rows of the standing crops.

The same rotation is not suited equally to every kind of soil. On the sandy soils of New England, abundant in silica, Indian corn, rye and the grasses naturally occupy more space than they would in a soil rich in lime. In such a soil as the last, wheat might take the place of rye and of Indian corn.

720. The farmer must find out, from the experience of others or from his own observation, what course is best for the particular soil he cultivates, and the particular object he has in view.

One may choose to keep sheep, another, only cattle for the market, another, cows for the dairy. A farmer living near a large market would pursue a course very different from one at a great distance. He would naturally make his farm resemble a large market garden.

721. **Other Rotations.** Usually a field laid down to grass may be profitably kept for mowing for several years. This being understood, and also that grass seed may often be conveniently sown with clover, either of the following may be an advantageous course:—

I.—1, corn; 2, beets; 3, rye; 4, clover; 5, grass.

II.—1, potatoes; 2, corn; 3, carrots; 4, cabbages; 5, beets; 6, clover and grass.

III.—1, potatoes; 2, beets; 3, beans; 4, cabbages; 5, parsnips; 6, corn; 7, clover.

IV.—1, tomatoes; 2, squashes; 3, carrots; 4, pease or beans; 5, cabbages; 6, clover.

V.—1, turnips; 2, parsnips; 3, corn; 4, potatoes; 5, rye; 6, clover; 7, grass.

For a long course, 1, potatoes; 2, beets; 3, squashes or melons; 4, carrots; 5, wheat; 6, parsnips; 7, rye; 8, turnips; 9, buckwheat; 10, corn; 11, clover; 12, grass.

It is found by experience that corn does not well follow turnips.

722. The famous Norfolk, (Eng.,) rotation is 1, turnips; 2, barley; 3, clover; 4, wheat. A favorite rotation in France is for the

1st year, winter wheat, 20 acres.

2d year, beets, carrots, potatoes, 10 acres; poppy or flax, 5 acres; colza, 5 acres.

3d year, oats and spring wheat, 10 acres; fall wheat, 5 acres; turnips, 5 acres.

4th year, clover or leguminous vegetables, 20 acres.

Poppies and colza are a special object of cultivation in France. From the seeds of both oil is made for light and for culinary use.

On rich clayey soils in England, a course which has been much used is 1, oats; 2, rape, for oil; 3, beans; 4, wheat sown with clover; 5 and 6, clover; 7, wheat; 8, rape. In rich loams, 1, oats; 2, turnips; 3, wheat or barley; 4, beans; 5, wheat; 6, fallow or turnips; 7, wheat or barley and grass seeds. But it must be remembered that the climate of England does not ripen Indian corn.

723. Rotation of crops is not indispensable. It may be the best economy, on the whole, of manure, of time and of labor. But the farmer who knows the precise use and value of the several mineral and other manures, may substitute, for a rotation of crops, a rotation of manures, which will enable him to grow, on the same field, again and again, the crop which may be most profitable for him or most in demand in his market.

CHAPTER XXII.

THE HARVEST.

724. The hay crop is usually the first of the harvest that requires attention. Before he can determine the proper time for mowing, the farmer must consider for what purposes his hay is to be used—whether he is to feed cows in milk, horses and working oxen, or young stock with it.

725. If it be used for feeding milch cows, it should be cut earlier than if it is intended for some other kinds

of stock, and at such a time and in such a manner as to preserve its juiciness and leave it as much like the green grass of the pasture as possible.

726. If it is to be fed to cows in milk, and the farmer wishes to get the greatest *quantity* of milk, grass should be cut just before coming into blossom. It is then most juicy, and will therefore produce a greater flow of milk than if allowed to stand longer. If the object is to secure the best *quality* of milk, with less regard to quantity, it may be cut in the blossom.

727. For feeding to store cattle, the grasses may be cut when in full blossom; for horses at work and for fattening cattle, it is better just after it has passed out of the blossom, or when the seed is said to be in the milk.

728. Grasses attain their full development at the time of flowering, and then contain the largest quantity of soluble materials, such as starch, gum, and sugar; these, with the nitrogenous compounds which are also most abundant at this time, are of the highest value for supplying nutriment to animals.

729. After flowering, and as the seed forms and ripens, the starch, sugar, &c., are gradually changed into woody fibre, which is nearly insoluble and innutritious.

730. This fact is well established, and shows that grasses in general should not be allowed to stand after the time of flowering. There is, indeed, a great deal of nourishment in the ripe seed; but not enough to make up for the loss in the stalk and leaves, if the mowing is put off till the seed is ripe. Grasses fully ripe will make hay little better than straw.

731. Grass is cut either by hand with the common scythe, or by the mowing machine, (Fig. 36.) With the

former, a good mower will go over an acre a day. With the latter, on smooth land, two horses and one man will mow at the rate of an acre an hour, or from ten to twelve acres a day, without over-exertion.

Fig. 36.

732. Besides mowing so much faster, the machine also spreads the grass evenly, saving the labor of spreading by hand. It also enables the farmer to cut all his grass nearer the proper time, and he is not obliged to let a part of it stand till it is too ripe.

733. After being cut, the grass should be frequently spread and turned, so as to dry as rapidly and as uniformly as possible. This may be done by hand with a common fork, or by a machine called a hay-tedder, a light revolving cylinder set with tines and drawn by one horse, by means of which the grass may be constantly stirred and kept in motion, and much time and labor may be saved.

734. When grass is partially or wholly cured, it may be raked by hand, or by a horse-rake, (Fig 37.) Raking

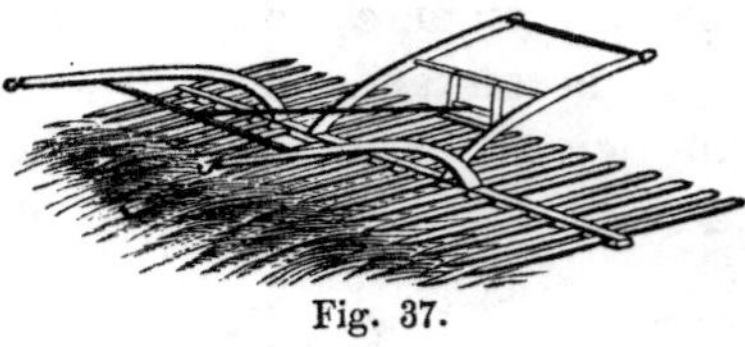
Fig. 37.

by hand is easy but slow, and thrifty farmers now generally use the horse-rake whenever they can. With the horse-rake, one man and horse can do as much work as ten men can in the same time without it. Hay cut in the forenoon should be raked before night, that it may not be exposed to the dews.

735. The time required for curing hay depends partly on its ripeness when cut, and much on the state of the weather. In good weather, if machinery is used, it may be cut in the morning after the dew has risen, and dried so as to be put into light cocks early in the afternoon, or before the dews of evening. A slight opening to the sun for an hour or two the next day should dry it enough, if it was cut while in blossom or before. Hay should be got in during the heat of the day.

736. Grass cured rapidly and with the least exposure, is more nutritious than that cured more slowly and longer exposed to the sun. If dried too much, it contains more useless woody fibre and less nutriment. The more succulent and juicy the hay, the more it is relished by cattle.

737. After the grass has been cut at the proper time, the true art of haymaking consists in curing it just enough to make it fit for storing away, and no more. The loss of the nutritive substances, which make the hay most valuable, is then stopped at the earliest moment. It is as great a mistake to dry grass too much, as to let it stand too long before cutting.

738. If the hay has not been perfectly dried, and there is danger that it may heat in the mow, it is well to have

alternate layers of the new hay and straw or old hay. In this way the heating may be prevented, and the straw or old hay will be so far flavored and improved, as to be relished by stock of all kinds. If there is much reason for apprehension, four quarts of salt to the ton may be sprinkled in.

739. Experience has shown that hay properly dried is not likely to be injured by its own juices alone; if it has been exposed to rain, it should never be put into the mow until it has been thoroughly dried.

740. Clover should be cut immediately after blossoming and before the seed is formed. It should be cured in such a manner as to lose as little of its foliage as possible, and therefore cannot be treated exactly as the natural grasses are. It should not be long exposed to the scorching sun, but after being wilted and partially dried, it should be forked up into cocks and left to cure in this position. The fourth or fifth day, when the weather is fair and warm, open and air it an hour or two, and it will then be fit to cart to the barn.

741. Clover cured in this way without loss of its foliage, is better for milch cows and for sheep than any other hay. It may also be fed to horses that are not hard worked, or to young stock, but it is most valuable for cows in milk. For other farm stock it is worth from two-thirds to three-fourths as much as the best hay.

742. If there is reason to fear that it is not sufficiently cured when stored away, it may be mixed with old meadow or swale hay or straw, putting first a layer of hay or straw, and then one of clover. Stored in this manner, cattle will eat both the hay and clover very greedily.

743. Lucerne should be cut as soon as it begins to flower, or even earlier. If allowed to stand later, it

becomes coarse and hard with much woody fibre, and is less relished by cattle. It is cured and used like clover.

744. The proper time to cut both wheat and rye is when the straw begins to whiten and shrink just below the head. This change will commence a week or more before they are fully ripe, and shows that the grain has ceased to receive nourishment from the roots. If taken in before getting dead ripe, it makes more and whiter flour, and the waste from shelling out is avoided. Wheat may be cut with the sickle, with the cradle, (Fig. 38,) or by the reaping machine, very similar in appearance to the mowing machine. A reaper in operation is shown in figure 39.

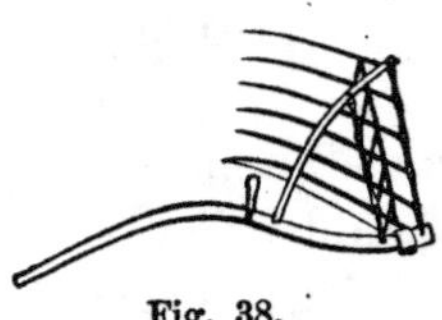
Fig. 38.

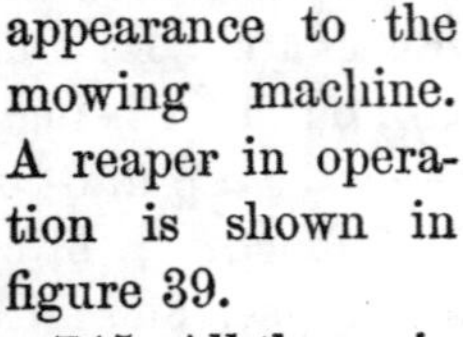

Fig. 39.

745. All the grain crops may be cut in the same manner, but oats and barley are most commonly mown and dried somewhat like hay, while the other grains are more frequently cradled or cut with the machine.

746. Indian corn should be gathered when the ears are glazed, but not perfectly hard. It is customary in many parts of New England, to cut the tops above the ears a little before this time, and when the stalks are still rather green. The corn is afterwards cut up near the ground, and taken to the barn to be husked. In other sections

the practice of cutting up at the ground and stooking, prevails. This is done after the kernel has become glazed, yielding but little juice when broken open, and when the leaves have begun to turn, but are still green. This practice saves labor, and adds to the quantity of fodder, and preserves its nutritive qualities better.

747. Potatoes should usually be dug in October. They may be thrown out by a furrow of the common plough, or with the spade or hoe, but the eight-pronged manure fork is better than either. They are liable to be injured by lying in the sun after they are dug, and if exposed to its direct rays are apt to lose their mealiness. But if kept in the shade until they are put into the cellar, they continue mealy much longer.

748. If the tuber of the potato while growing is exposed to the light and air by lying near or on the surface, it becomes disagreeable to the taste, green and waxy, and sometimes even poisonous, and when cooked will be found to be soggy. The effect produced on potatoes lying in the sun after digging is a little like this, though much less in degree, perhaps, on account of the shorter time they are exposed. But seed potatoes may be exposed to the sun before planting with great benefit.

749. The harvesting of turnips should be commenced in the early part of November. The Swede or ruta-baga, may be lifted out of the ground, the tops cut and the roots stored in a cool, airy cellar. The tap-root may be cut off to prevent sprouting in the cellar.

750. Carrots may remain in the ground till the late hard frosts, or till the early part of November. They may first be topped in the ground by running a sharp hoe or knife, (Fig. 40,) along the rows, and then may be raised with the common hand-fork, or a deep furrow may

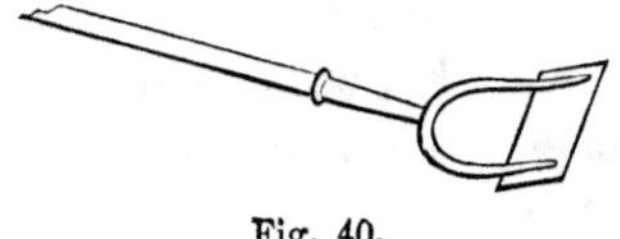

Fig. 40.

be made by the plough running as near as practicable to each row, after which they may be easily thrown out. After drying sufficiently, they are ready to be removed to the cellar. Parsnips may be taken up in the same way, or a part of the crop may be left in the ground till they are wanted to use, in spring.

751. Mangolds should be pulled and stored with as little bruising as possible. The least injury will sometimes cause them to decay. If properly harvested, this root keeps well till late into spring.

CHAPTER XXIII.

DISEASES AND ENEMIES OF GROWING PLANTS.

752. Disease is the result of deranged vital action. It is brought about both by predisposing and by exciting causes.

753. Whatever diminishes the natural vigor of the plant, but does not of itself produce a specific form of disease, as excessive stimulation, want of proper nourishment, and the propagation of any species for many years without mixing with other varieties of the same species, or in common language, not changing the kind of seed planted, is a predisposing cause.

754. An exciting cause of disease is one which acts suddenly upon the previously debilitated plant, and

produces such a change in its vital action as to excite a distinct form of disease, as sudden changes of temperature, or of the electrical condition of the atmosphere, hot and damp weather at unusual periods of the season; also, sometimes, mechanical violence.

[NOTE.—Fungi and insects have, by some, been considered as exciting causes of disease, while others regard them as resulting from previously existing disease.]

755. Among the diseases in which parasitic plants appear, or which are caused by parasitic plants, may be mentioned mildew, blight or red rust, smut and ergot.

756. The term mildew, or *meal dew*, is most properly applied to the mould or fungous growth on the leaves of trees and some forage plants, in the shape of white mealy patches.

757. But it is most commonly applied to a disease in wheat and barley, also called rust. It appears in the shape of small spots of dingy white, oblong in shape, showing itself first on the upper side of the leaf, but soon on the lower side and the stem also. The white mildew attacks many species of plants, especially roses, peaches, hops, vines, pease, the maple tree, &c.

758. It is first seen as round white or yellowish mealy spots, composed of very delicate creeping threads. As the disease develops, these spots throw off spores or cells, which attach themselves to other plants and produce similar fungi or spots on them. There is no other change in the appearance of this disease, and no change in color. The oidium of the vine is a kind of white mildew.

759. The simplest and most effectual remedy for this is to take a lump of stone lime of two or three pounds weight, and about the same quantity of sulphur. Pour hot water on them, which, by the slacking of the lime,

causes both to dissolve readily in water. This is sufficient for a barrel of water, and when used may be filled up again. It may be applied with a syringe or sprinkling pot to the foliage of affected plants. Sulphur vapor is also a certain remedy for mildew for plants or vines under glass.

760. The wheat mildew is very different in its nature from that found on trees or vines, which may be called the white mildew, though its effects are somewhat similar.

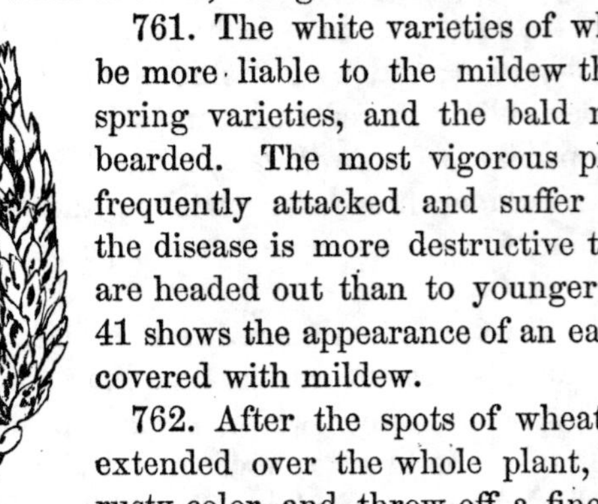
Fig. 41.

761. The white varieties of wheat appear to be more liable to the mildew than the red or spring varieties, and the bald more than the bearded. The most vigorous plants are most frequently attacked and suffer the most, and the disease is more destructive to plants which are headed out than to younger ones. Figure 41 shows the appearance of an ear of ripe wheat covered with mildew.

762. After the spots of wheat mildew have extended over the whole plant, they assume a rusty color, and throw off a fine dust which is yellowish at first, but soon turns brown and rusty by exposure. Hence the disease often goes by the name of rust.

763. Wheat growing on low, undrained lands, with a peaty or calcareous soil, is most liable to be attacked by mildew, but this disease often appears on sandy soils and on the stiffest clays, especially when a few days of damp, foggy weather, are followed by a hot sunshine.

764. No remedy is known which can be relied on to protect against this mildew, but the free use of salt or saline manures, soaking the seed in brine, or sprinkling the plants with salt dissolved in water at the rate of half a pound to the gallon, are the most effectual.

765. Salt or brine thus used should be applied on a cloudy day, or just at evening. The solution of salt acts almost instantaneously, where it touches the parts affected with rust.

766. Smut is a disease which attacks Indian corn, wheat, barley, oats, millet, and some other kinds of grasses. It derives its name from the fact that where it exists the receptacle of the seed is filled by a dark, sooty or dusty mass caused by an internal parasitic fungus. The first stage of smut in an ear of oats is shown in figure 42.

767. The farina or mealy substance of the grain affected by this disease is decomposed, and the whole grain and the husk are covered with the black powder, and are often swollen to a very large size.

Fig. 42.

768. Smut, like mildew, prevails in every variety of soil and in all localities and countries. But hot and moist climates are more favorable to its development than cold and dry ones.

769. It less frequently attacks wheat, than corn and oats. Its presence in wheat may be known by the blackish color of the ear, or before the ear has burst from its sheath, by yellow spots which appear on the upper leaf, and the drying up of the point or end of this leaf. An ear of wheat partly sound, and partly covered with smut, is shown in figure 43, and another wholly covered with smut, and dried up, in figure 44. In oats the diseased plants are of a paler green, and generally smaller than the rest. A head of oats wholly covered with smut, is shown in figure 45, and an ear of barley completely smutted and dried up, in 46.

Fig. 43. Fig. 44. Fig. 45. Fig. 46.

770. Quicklime, common salt, blue and white copperas, and many similar substances, have been used with success to prevent smut. A sort of solution or pickle is formed of one or more of these substances, as lime and salt, lime and Glauber's salt, salt and copperas, the object being to make a compound corrosive enough to destroy the parasitic fungus without destroying the grain. Very strong putrid urine, or the drainings of the stable, are sometimes sufficient.

771. Before soaking the seed in any pickle or brine formed of the above-named substances, it should be thoroughly washed and cleansed in pure water, taking care to remove all floating grains, and to pour off the water without allowing it to come in contact with other grain, so as to convey the disease by contagion.

772. There is a disease known by the name of blight, canker, smut-ball, pepper-brand, &c., which is often confounded with smut, but really very different. It has been supposed to be a fungus in the seeds of wheat, by means of which the farina was replaced by a whitish substance which finally became a fine powder, the outside or skin of the seed being untouched, and giving no signs of the presence of the disease. Figure 47 shows a section of a cankered grain of wheat.

Fig. 47.

773. But more recent investigations indicate that it is caused by microscopic animalcules or thread worms, which possess the remarkable power of remaining perfectly dry and hard for years, and then regaining life and motion when moistened.

774. Grain affected by this disease, becomes a hard shell filled with powder, which is usually white. This powder has no trace of starch, but is composed entirely of microscopic threads, which are stiff, dry, hard worms. When found in new grain, if placed in water, they show signs of life very quickly. If very old, it requires many hours or even days to revive them. Several thousand of these worms may be found in a single kernel.

775. When these diseased grains are sown with sound, the moisture gradually revives the worms. They break through the thin shell of their prison, and seek the young shoots of the wheat which has germinated, are carried up by the growth of the plant, or, if the weather be wet, by their own exertions, effect a lodgment in the young kernel, and lay their eggs there.

776. At the time of the ripening of the grain the parent worms are dead, the shells of the innumerable eggs which have produced larvæ have been absorbed, and

nothing is seen on breaking through the covering of the seed, but what appears to be an almost impalpable powder, each grain of which is a dry, hard, thread-like larva.

777. Threshing very easily breaks the thin shells which surround this powder, and it rises in the form of dust, causing severe smarting in the eyes, and some irritation of the throat and coughing, as the animalcules are set in motion by the moisture. No serious results follow, however, except that more or less of this dust attaches itself to sound kernels, thus propagating the disease.

778. Where the seed is supposed to be at all effected in this way, it should be thoroughly washed in clean water, several times renewed. All the grains that float should be carefully taken out. The seed may then be soaked in a brine or pickle much as follows:

779. For every two bushels of seed take three pounds of caustic lime in lumps, and sixteen pounds of Glauber's salts. Dissolve the latter in six or eight quarts of water, and whilst they are dissolving, slack the lime. Put the grain into a tub and stir well, pouring on the solution of Glauber's salt at the same time. Now sprinkle in the slacked lime, constantly stirring the seed until the whole is covered with lime.

780. The term blight is properly applied to a withering or blasting of the foliage, by whatever cause produced. It may be the result of sun-stroke or frost,—a plague of insects or fungi. It may be caused by drought, heat, cold, over-manuring or insufficient nourishment, or by an original want of vigor in the seed. Still it is blight. The term is also often used in this country as including mildew, rust, and many other affections of the kind to which plants are liable.

Fig. 48.

781. Ergot is a diseased growth which is quite common in rye and among our grasses. It appears in the form of a hard, brittle, blackish spur, of a form represented in figure 30, and takes the place of the healthy seed, though very much larger, being sometimes more than an inch in length. An ear of rye attacked by ergot is shown in figure 48.

782. Ergot has been supposed to be caused by a parasitic fungus growth starting from the ovule or rudimentary seed. Instead of sugar, albumen and the other substances of which sound grain is composed, this spur or morbid growth contains ammonia, considerable nitrogen, and an oily substance.

783. Ergot most frequently prevails on low, damp soils, in sheltered situations, but often on sandy soils, and sometimes on all varieties of soil.

784. There is no remedy for ergot after it has appeared, but it may be guarded against, to some extent, by thorough drainage and by carefully cleansing the seed, and, if necessary, picking it out by hand to avoid planting any that is diseased. If fed to some animals, it often produces very bad effects.

785. Trees, especially fruit trees, are often injured by pruning or grafting done unskilfully, or at the wrong season, or severe bruises inflicted in careless ploughing around them, or otherwise.

786. Fruit trees can be pruned with safety at any time except in March and April. Grafting is usually done in May or June. Both operations should be performed carefully.

787. When trees have been severely bruised, or large branches have been broken off by accident, the wound or

broken end should be well covered over with clay or with grafting wax. Many valuable trees might be saved from permanent injury or destruction in this way.

788. Some of the insects most injurious to vegetation are cut-worms, apple-tree caterpillars, canker worms, apple-tree borers, codling-moths, the curculio, the striped or cucumber-bug, the squash-bug, the onion-fly, the wheat midge, the chinch-bug and the army worm.

789. The cut-worms destroy many of our garden and field vegetables by eating off their tender stalks at the surface of the ground. They are the caterpillars of moths belonging to the night-flying division, one of which is represented in figure 49, and the cut-worm in figure 50.

Fig. 49.

Fig. 50.

790. If holes be made with an iron bar or smooth round stick near the roots of the plant, the worms will fall into them, and may be killed; they may also be found early in the day close to the roots of the plants they have cut down during the night.

791. Certain species of ground-beetles, and ichneumon-flies destroy great numbers of cut-worms, and similar caterpillars, and hence are very useful to the farmer, and should be recognized and spared on this account.

792. The ground-beetles are very active in their motions, and although varying greatly in size, more or less resemble in their general outline and conformation, figure 51, which is one of the largest of its class, and is commonly called the caterpillar-hunter.

793. Ichneumon-flies are of various species and dimensions, but they all have four wings of membranous texture, and the general appearance of a wasp. Some of them pierce the eggs of other insects and deposit their own within them; others insert them beneath the skin of a living caterpillar, where they hatch into little maggots, which devour its flesh and soon put an end to its life. Figure 52 represents a species (natural size and magnified) which deposits its eggs in the body of the common grape-vine caterpillar. Figure 53 shows the caterpillar after the maggots of the ichneumon have finished eating and, returning through the skin of the caterpillar, have spun their cocoons upon its surface.

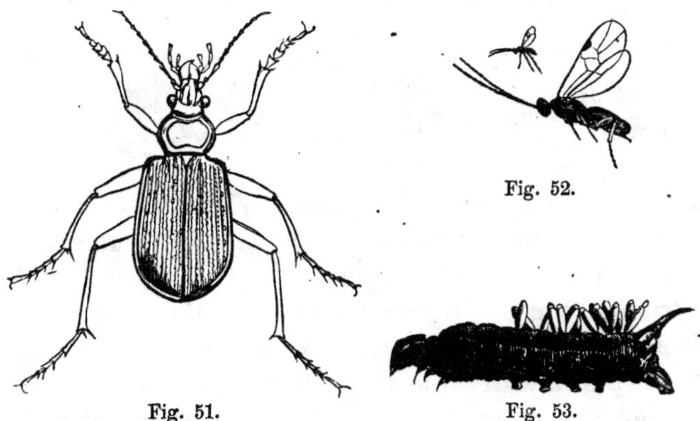

Fig. 52.

Fig. 51. Fig. 53.

794. The apple-tree caterpillar may be guarded against by carefully removing all the nests as soon as perceived, and crushing both larvæ and nests. If this practice be well followed up, they may be eradicated from a whole neighborhood. A round brush fixed to the end of a long pole is the most convenient instrument for reaching the nest. The eggs, (Fig. 54,) which are laid the previous season, may be

Fig. 54.

seen in the form of a small bracelet or broad ring around the slender twigs when the leaves have fallen from the trees. With a little observation these can be readily distinguished, and by means of a light ladder the twigs containing them may be reached, when they should be cut off and burned. This, if done any time during the winter, will save much trouble in the spring after they have hatched.

795. The best means of protecting trees against the canker-worm, (Fig. 55,) is by preventing the deposit of the egg. The wingless female (Fig. 56,) lays her eggs (Fig. 57, natural size and

Fig. 55.

Fig. 56.

Fig. 57.

magnified,) on the bark of the tree, and ascends the tree for this purpose during the warm days of winter and spring. A coating of tar on a strip of cloth round the trunk, frequently renewed during that time, will often prevent her ascent; or a little trough may be put round the tree filled with a mixture of tar and oil, enough oil being put in to keep it in a liquid state, or with the "bitter water" obtained in the manufacture of salt, which will have the same effect. This will neither freeze nor evaporate readily. The winged male is shown in figure 58.

Fig. 58.

796. The codling moth, (Fig. 59,) produces the small whitish worms that bore holes into the young unripe

apple and other fruit, and cause it to fall off. The windfalls should be picked up often and given to swine, or if convenient, the swine may be turned into the orchard to pick them up. The grub will thus be prevented from going into the ground. Old cloths may also be tied in the crotches of the limbs of fruit trees. The worms take refuge in them and may be killed.

Fig. 59.

797. The curculio, (Figs. 60 and 61, the small line between them showing the natural size,) does much injury, attacking the plum particularly. Fruit bitten by it may be distinguished by a little crescent-shaped mark, and should be collected and burned. If sheets be laid under the trees, and the trees then be shaken, the insects will fall into the sheets and may be put into hot water. If chickens in coops be kept under the trees in summer, they will destroy immense numbers, as do the small birds also; toads and bats too do good service in this way.

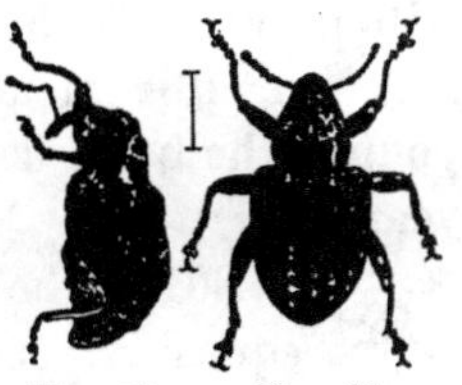
Fig. 60. Fig. 61.

798. The apple-tree borer, (Fig. 62,) with its larva, (Fig. 63,) is ruining many an orchard where his presence is not suspected, and trees should frequently be examined that it may be discovered as soon as possible. The borer enters the tree just at the surface of the ground, and by removing the soil and rubbing the bark with a coarse cloth after the first of September, the young insect may easily be destroyed.

Fig. 62.

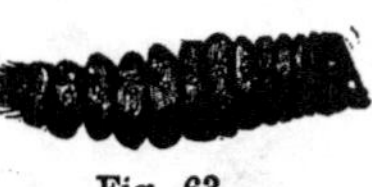
Fig. 63.

799. The eggs are hatched in July, so that the larvæ will have attained considerable size, and may easily be seen and dislodged without difficulty. Even later than this, careful examination will show that they are still near the surface, and may be reached by a slender piece of whalebone or wire, run into the new-made hole.

800. The chisel and the hammer must be used only when all other means fail. Washing with whale-oil soap will prevent the laying of the eggs, but it will not do to rely on this alone. If unmolested when still quite young, the borer continues his depredations from year to year.

Fig. 64.

801. The striped beetle, (Fig. 64,) attacks squashes, cucumbers, melons, and other plants. To prevent injury from it, the plants should be sprinkled as soon as they are up, with plaster of Paris or slacked lime put on in the middle of the day, or they may be covered over with coarse millinet or lace, which answers quite as well. If squashes or cucumbers are not planted till the 10th of June, they will usually escape the attacks of this insect.

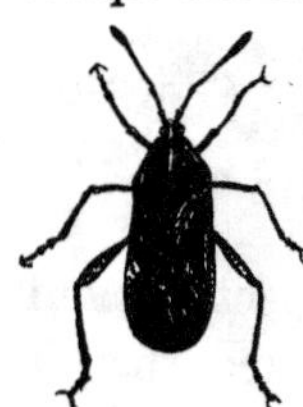

Fig. 65.

802. Squash bugs, (Fig. 65,) may be destroyed by placing shingles on the ground round the vines, and killing the bugs which will be found in the morning collected on the under side of them.

Fig. 67. Fig. 66.

803. The onion maggot pierces the centre of the onion and kills it, the egg from which the maggot proceeds being laid near the root by the onion fly, (Fig. 66.) The pupa of this insect is shown in figure 67. The use of soot in the drills is the best preventive known.

804. The wheat midge, (Fig. 68, magnified, the small mark at the left shows the natural size,) is itself exposed to the attacks of other insects. An ichneumon fly deposits its eggs in the larvæ of the midge, and the larvæ hatched from them prey upon the body on which they find themselves. Many are thus destroyed. If the stubble be collected and burned, innumerable grubs of the midge will be consumed, and the good work of the ichneumon be aided.

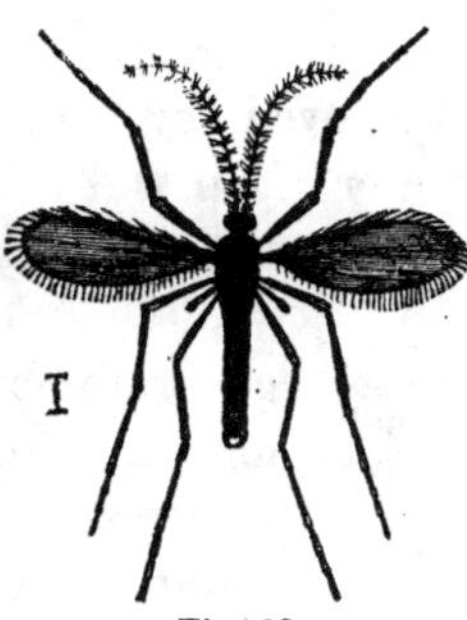

Fig. 68.

805. The dor bug, as it is called, (Fig. 69,) is properly a beetle, and the parent of those large white grubs which feed upon the roots of grass and grain, and are so frequently turned up by the spade or plough. Domestic fowls devour great numbers of them in the latter state, and many of the beetles themselves are eaten by skunks and weasels.

Fig. 69.

806. The locust tree borer, (Fig. 70,) which ruins so many of the finest trees, is the caterpillar of a moth,

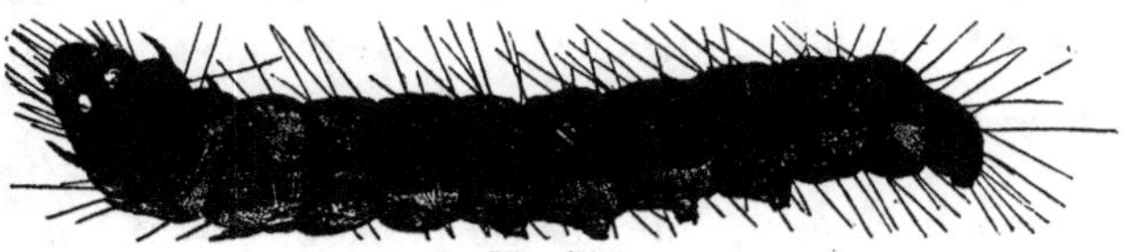

Fig. 70.

which deposits its eggs in the deepest clefts of the bark. They hatch into grubs which commence boring into the very heart of the tree, piercing and mining it with their burrows for three years before they make their appearance in the winged form. In this state it is the most easily destroyed, by hanging wide-mouthed vials of sweetened water upon the trees, which attract not only these, but also many other noxious insects.

807. Figure 71, the rose-bug as it is very generally but improperly called, is a *beetle* belonging to the chafer family, and is very destructive to flowers and foliage. When it occurs in great numbers upon bushes that can be reached by hand, it should be shaken off into pans of hot water. The larva lives in the ground like the others of this family, and when turned up by the plough is greedily devoured by poultry.

Fig. 71.

Fig. 72.

808. The common click-beetle or spring-beetle, (Fig. 72,) is the parent of the wire-worm, and should be killed whenever met with, as well as all of this family, which can be rèadily distinguished by their faculty of springing into the air when laid upon their backs, by means of a peculiar joint beneath the thorax.

Fig. 73.

809. The striped potato-beetle, (Fig. 73,) is often found eating holes through the leaves, in both the perfect stage and in the larva, which is the filthy slug so common and so injurious in some seasons. Lime or ashes sprinkled profusely upon the plants, will often destroy them, and when this fails, they can be shaken into dishes of boiling water or salt and water.

810. The oak-pruner, (Fig. 74,) is the parent of a white grub, (Fig. 75,) which bores into the small branches and twigs of the oak tree, making a cylindrical burrow, and cutting the branch nearly through; after which it retires

toward the end and changing into a pupa, (Fig. 76,) falls to the ground with the branch which is torn off by the wind, and remains till spring, when it emerges a perfect beetle, like the parent. To prevent its ravages, the branches found beneath the trees in the fall and winter should be collected and burned.

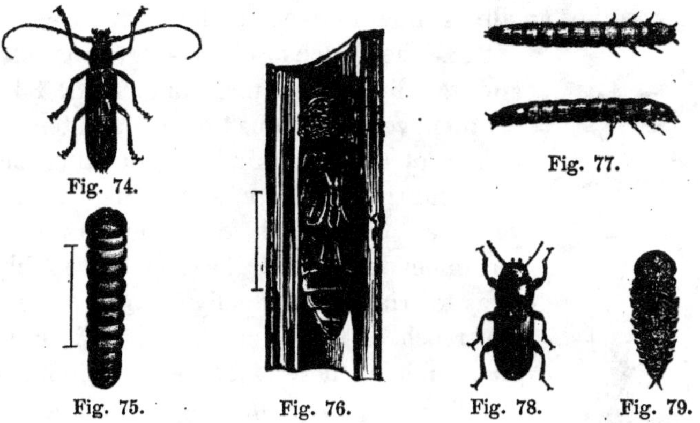

Fig. 74. Fig. 75. Fig. 76. Fig. 77. Fig. 78. Fig. 79.

811. The meal-worm, (Fig. 77,) which is found in meal chests, is hatched from eggs deposited by a common beetle, (Fig. 78,) which can be attracted in great numbers by a light in the evening, or moist meal exposed to the air, and should be killed wherever found. Figure 79 represents the pupa of the same.

812. Apple and pear trees are sometimes covered with small scales, as in figure 80, which represents those of the apple, natural size and magnified. A solution of potash, not too strong, or whale oil soap suds applied with a stiff brush, will speedily remove them. These insects belong to a very numerous class which vary greatly

Fig. 80.

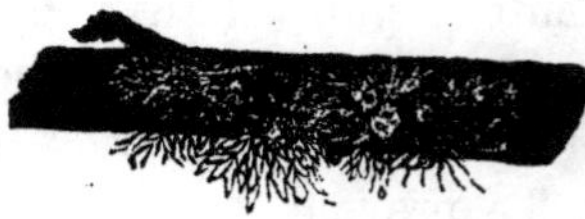
Fig. 81.

in their appearance; some are covered with a white flocculent matter so as to entirely conceal their bodies, as in figure 81, and others are entirely naked, and of various forms. The application of strong soap-suds, will be an almost infallible remedy for all these cases.

Fig. 82.

Fig. 83.

813. The chinch or chintz-bug, (Fig. 82,) and the little-lined plant-bug, (Fig. 83,) are often very injurious to green and tender plants, of different kinds, by sucking the sap from them. No effectual remedy has yet been discovered for them, but it is recommended to water the crops thoroughly so as to bring them rapidly forward beyond the reach of these insects. This is only practicable upon a small scale. Wild birds and domestic fowls destroy an incredible amount of these and other similar pests annually, and we must take especial care of the former, and allow no person to kill or molest them upon our premises, if we would have our crops secured from their numerous insect enemies.

Fig. 84.

814. The army-worm, as it is called, (Fig. 84,) is the caterpillar of a common night-flying moth, (Fig. 85,) and is found in meadows, devouring the blades of grass, and occasionally in corn and grain fields. It has many enemies, in the form of ichneumon-flies, and other parasites, and is eaten by many birds. It is rare that so many escape destruction by these means as to prove seriously dangerous to the whole crop in any place. But when they make their

appearance in unusual numbers, they can be checked by digging in their path deep trenches, with perpendicular sides, into which they will fall and may be disposed of. After they reach their full size they suddenly disappear, and may be found an inch or two below the surface of the ground in the shape of a mahogany-colored pupa, (Fig. 86.) After remaining in this state about a fortnight, they emerge in the moth form, and may be killed by building fires after dark about the fields that were injured by them, into which they will fly, or by suspending on the trees in the vicinity, wide-mouthed bottles of sweetened water.

Fig. 85.

Fig. 86.

815. The plant-lice are a numerous family, and often very injurious to young shoots, by sucking the sap and causing the plant to wither. They are found usually in clusters, with and without wings, and may be distinguished by their rounded bodies, slender legs, and delicate pointed beak, which is bent underneath the body when not in use. Figure 87 is a good representation of the male of one of the most common species. In some seasons vast numbers may be found collected upon the heads of wheat, oats, and other grain crops, and by depriving the fruit of its requisite amount of nourishment, they cause it to shrivel and ripen prematurely.*

Fig. 87.

* For the most exact and valuable information upon Insects Injurious to Vegetation, reference should be made to the superbly illustrated edition of Dr. Harris' treatise on Insects, just published. No farmer's library is complete without it, and it ought to be kept for reference in every school-room.

CHAPTER XXIV.

MANAGEMENT OF FARM STOCK.

816. The stock of the farm consists of horned cattle, horses, sheep, swine and poultry.

817. Horned cattle are kept chiefly for their milk, their labor, and for the production of beef. They also consume and thus make useful many products of the farm which would otherwise be lost, and furnish manure for the enrichment of the soil.

818. They are divided into certain races, breeds or families, distinguished by different qualities or characteristics which have been produced or developed by varieties of climate and soil, and by the manner in which they have been required to live by man.

819. There are five distinct races or breeds in this country, known as Ayrshires, Jerseys, Short-horns, Devons and Herefords. Individuals of other breeds have been imported from time to time, but their number has been so small that they have had little effect on the stock of the country.

820. No one of these breeds unites, in a very high degree, all desirable qualities. Some are best fitted for giving milk, others for beef or labor. Cattle should therefore be selected with regard to the specific object for which they are wanted, and that object should be had in view in their keeping.

821. The Ayrshires take their name from the county of Ayr, in Scotland, where they originated seventy or eighty years ago. They are kept chiefly for dairy

purposes, for which they are admirably adapted, on account of the large quantity of milk they give in proportion to their size and the amount of food consumed. Their milk is of good quality, though not, usually, so rich in butter qualities as that of the Jerseys or the Devons. They are well adapted both for beef and for labor, though in these qualities they are probably surpassed by the Devons, or the Herefords.

822. The Jerseys are celebrated for the richness of their milk, and the excellence of butter made from it. They came from the islands of Jersey and Guernsey, in the British Channel, where they have been highly valued for dairy qualities for many years. They are ill-adapted for labor, and their beef-producing qualities do not compare very favorably with those of some other breeds, although they are easily fattened, and their flesh is of good quality.

823. The improved Short-horns are large in size, and, in a rich and fertile section of country, are well-adapted for the production of beef. They come to maturity at an earlier age than any other family of neat cattle, and attain a greater weight.

824. They first became known in the luxuriant valley of the river Tees, England, and first really celebrated in the neighborhood of Durham. Hence they were for many years called Durhams or Teeswaters. They have been extensively introduced into this country, and have had a great influence upon our stock.

825. The North Devons are remarkable for great uniformity of color and size, and are kept chiefly for beef and as working cattle. They come from Devonshire, in the southern part of England. They are small, hardy, and easily adapt themselves to short pastures. Their milk is rich in quality, but deficient in quantity.

826. The Herefords, so-called from the county of Hereford in England, where they originated, have nearly the same qualities as the North Devons, but their size is considerably larger. They are kept mainly for their beef, which is of peculiar excellence.

827. These are the distinct breeds. The common stock of the country, often called Natives, does not constitute a fixed breed or race. It consists of a mixture of most of the established races, and is extremely variable in its qualities. Animals might be selected as good, or perhaps better than could be found among the well-marked families, and as working oxen, they generally excel, but as a whole, they are not to be depended upon for any uniformity of qualities.

828. Only good stock should be kept on the farm. It costs no more to keep a good animal than an inferior one. One that will scarcely pay the cost of rearing and feeding, will require about as much care and food as another which will pay a large profit.

829. Success in raising stock will depend very much on its management when young. If it be not then well cared for, and supplied with sufficient and proper food, the grown animal will be of poor quality, whatever the breed may be.

830. All animals require nutriment in some proportion to their live weight, those which are still young and growing, needing more in proportion than those already arrived at maturity.

831. A full-grown animal requires only food enough to supply the daily waste of the system. One that is growing must have enough to supply the daily waste, and to meet the additional demand for nutriment arising from its constant increase in size and weight.

832. For these reasons, young animals should have greater care, better shelter, and more generous feed than they commonly do. Yet they should not be overfed; they should receive enough to keep them growing thriftily up to the time of their maturity, and the necessary quantity must be determined, to some extent, by observation in each case, though general rules are sometimes laid down, fixing the proportion of food required at certain ages.

833. Farmers are too apt to consider how they can get their cows through the winter with the least possible food, taking no care to prepare them for the giving of milk abundantly in the spring.

834. In consequence, cows often come out in spring reduced in flesh and in blood, and have hard work to make up their loss by means of the food which would otherwise have gone to the production of milk.

835. The less cows in milk are exposed to the colds of winter, the better. They eat less, thrive better, and give more milk, when housed all the time during extreme cold weather. In stormy weather it is good economy to water them in the stall, rather than turn them out to seek water in the yard.

836. In the care of cattle, regularity is of the highest importance, especially in feeding. A regular system of feeding, milking and cleansing the stables, should be strictly adhered to.

837. Cows give a greater quantity of milk in winter, if fed on moist and succulent food. If hay, cornstalks, straw and other similar substances fed out to them, are moistened with warm water and then allowed to stand a few hours in this condition, they are rendered more nutritive.

838. When the object is to obtain the greatest quantity of milk, cows should have rich, juicy grass or clover,

brewers' grains, warm mashes, turnips, or other roots containing a great deal of water; they will also do better for whey, if at hand, and should have as much water as they will drink.

839. But if a rich milk be desired, they should be kept on drier food, such as clover, hay, Indian meal, shorts, oil cake ground into meal, and some roots. Oats and barley meal are good, but are generally too expensive.

840. When cheese is to be made from the milk, ground beans, or pease and clover with some oil meal, are better. They make the milk very rich in curd, as they contain a large amount of gluten, which is nearly the same as the curd of milk.

841. The manner of milking exerts a powerful influence on the productiveness of the cow. A slow and careless milker, or one who treats her harshly, soon dries up the best of cows. The animal must be approached gently, never struck or abused, and the operation of milking begin gradually, steadily increasing in rapidity, until all is drawn. If the milking is performed in the stall, it is a good plan to feed at the same time with roots or some other palatable food.

842. If the object be to raise beef, a close built, round and compact form, with small bones and round muscles should be sought. Animals thus shaped require less food and fatten more easily, than those of heavy, bony frame and flat muscles.

843. When fattening, animals should be kept quiet and warm, and fed on fatty or oily food, such as oil meal, Indian meal, good hay and turnips. A moderately dark stall conduces to quiet and promotes fattening.

844. To ascertain the results of feeding under various circumstances, the most careful experiments were made

upon sheep, by selecting those of nearly equal weight, and feeding for four months under the following conditions. One was wholly unsheltered, another in an open shed, and another in a close shed and in the dark. The food was alike, one pound of oats each per day, and as many turnips as they would eat. The first consumed nineteen hundred and twelve pounds of turnips, the second thirteen hundred and ninety-four pounds, and the third eight hundred and eighty-six pounds, or less than half of those eaten by the first. The first gained twenty-three and one-half pounds in weight, the second twenty-seven and one-half pounds, and the third twenty-eight and one-fourth pounds. For every one hundred pounds of turnips eaten, the first gained in weight one and one-eighth pounds, the second two pounds, and the third three and one-sixteenth pounds. The one confined in the dark ate less than half as much, and gained more than the unsheltered one.

845. If the farmer wish to make as much manure as possible from a certain quantity of hay, straw or turnips, the stock should be kept in a cool place where the external air is not entirely excluded, and allowed to take a great deal of exercise. If fed on rich food, like oil or Indian meal, the manure of the animal is of far greater value.

846. In general it may be stated that food which has been crushed, ground or cooked, is more easily and completely digested by stock, and furnishes more nourishment. Three pounds of ground corn are equal to about four of unground, and three of cooked Indian meal, to about four of the same meal uncooked. Meal and roots are usually cooked by boiling.

847. But where animals are already fattened, it is found to be better to keep them on dry, hard food for a few days before sending them to the butcher, as the fat is thus

made harder, and the meat is more readily salted through, keeps better, and shrinks less in cooking.

848. An animal in good condition will usually lose from thirty-two to forty per cent. of its live weight in dressing. If very fat and well formed, the loss will be about one-third, or thirty-three per cent. In a fat sheep, on an average, it will be from thirty-five to forty-five per cent.

849. Working cattle should have strength, docility and quickness of action. Strength lies in the muscles and tendons. Docility is commonly the result of good training. Activity is to some extent the result of breeding, and certain races, like the North Devons, are remarkable for this quality.

850. In most cases oxen are to be preferred to horses for common farm labor. They are more easily raised, become more valuable as they gain in size, weight and condition, and may be sold for beef when no longer fit for work. The harness used for them is cheap, and they are better adapted to slow and heavy work, especially on rough farms. Horses work faster, and are sometimes more profitable on easily tilled farms.

851. Horses are classified, according to the uses to which they are put, into roadsters, or horses of general utility, farm or draught-horses, and thoroughbreds or racers, used mostly for sporting purposes.

852. The horse requires a light and well-ventilated stable. If he stand much in a dark stall, his eyes are often so affected as to be irritated when he is brought into a strong light. In this way horses are frequently made skittish and unsafe.

853. The horse should, from the first, be treated with great gentleness, often led about by the halter long before

he is old enough for the harness, and made to feel that his master is his friend. Kind treatment will do much to insure docility, and greatly enhance the value of the animal for all practical purposes.

854. Well-lighted barns and stables do much for the general health and vigor of the animal system, and a full supply of pure fresh air is as essential as food. Especially is this the case for horses.

855. But animals should not be exposed to currents of air in the stalls. A chimney-shaped box opening near the floor inside, and carried up and out under the eaves, is thought to be a good mode of creating an outward draught and purifying the air.

856. The temperature of stables should be moderate, neither very warm nor very cold. Great warmth in them is unhealthy, and a considerable degree of cold makes a larger quantity of food necessary to keep up the natural animal heat.

857. All animals should be treated with constant kindness. Nothing is so likely to overcome viciousness. The horse, especially, is very sensitive, and if always gently handled, will give his owner far less trouble, and will be more easily managed and much more useful.

858. There are several breeds of sheep, the best being the South Downs and Cotswolds, which are generally sold to the butcher for mutton or lamb, and the Merino which furnishes the best wool. The Leicester sheep was very highly prized at one time, and this breed or grades which are known by the name of Leicesters, is thought well of still, but the Cotswolds and the Downs have, to a considerable extent, taken their place in localities where sheep are raised for the butcher.

859. In the vicinity of large markets, and where pasturage is expensive, it will be found to be most profitable to raise sheep for the market, only making wool a secondary object. But in remote and mountainous regions, where land is cheap and not suited to cultivation, they may be profitably kept for the wool. Many, however, think that even for wool, the larger breeds may be equally profitable, on account of the greater weight of their long and coarse wool, which is well suited for many kinds of fabrics, and commands a good price in the market.

860. Mutton of a choice quality, usually brings a higher price in the market than beef, though it costs much less pound for pound to produce, and the offal or waste is less. The objection to keeping the smaller breeds or the old natives, based on the expense of fences, does not apply so strongly to the larger or mutton breeds, like the Cotswolds, which are generally very quiet and easily kept.

861. One of the most important matters to be attended to in the keeping of sheep, is their shelter in winter. They require less food, and do better when well protected, than when exposed. Good ventilation is also very important, hence it is best to give them sheds open to the south.

862. To ascertain the difference in the cost and gain of proper shelter, and exposure to the weather, for sheep, in the milder climate of England, twenty were kept in the open field, and twenty others of nearly equal weights were kept under a comfortable shed. They were fed alike for the three winter months, each having one-half pound of linseed cake, one-half pint of barley, and a little hay and salt per day, and as many turnips as they would eat. The sheep in the field eat all the barley and oil cake, and about nineteen pounds of turnips each per day, as long as the trial lasted, and increased in all five hundred

and twelve pounds. Those under the shed consumed at first as much food as the others, but after the third week they each ate two pounds less of turnips per day, and in the ninth week two pounds less again or only fifteen pounds per day. Of the linseed cake they also ate about one-third less than the other lot, and yet increased in weight seven hundred and ninety pounds, or two hundred and seventy-eight pounds more than the others.

863. The winter feed of sheep should include a proper portion of green and succulent food, in addition to fine hay or early cut clover. Unless it be of good quality, much of it is rejected and wasted.

864. Ten fine-woolled or Merino sheep, will eat about as much as a medium-sized cow. The larger sheep consume more. The Merinos yield the best wool, the Cotswolds the most wool and mutton, and the South Downs mutton of the best quality.

865. It will be found useful to attach bells to several of the flock. By this means dogs may often be prevented from attacking them, and if the sheep are molested a warning is given. This is also a protection against foxes.

866. There are many breeds of swine, as the Suffolk, the Essex, the Berkshire, the Chester, &c., each of which has its peculiar excellence, but the more common distinction is into large and small breeds. The choice must depend much on thriftiness and early maturity, or a disposition to fatten readily, for on these qualities will depend largely the profit to be derived from keeping them.

867. The food of swine may be a little sour, without injury, if it does not stand till a strong fermentation takes place; indeed, more pork will be obtained when green vegetables, meal and potatoes, are boiled and allowed to

become sour before feeding them out, than if given while still sweet.

868. Poultry may be kept to a limited extent about the farm house, with a large profit on the outlay, if judiciously managed. The attempts to keep large numbers of fowls together with an idea that if a few are profitable, a large number must be profitable in proportion, have generally failed.

869. To be of any profit in winter, fowls require a supply of animal food. This they obtain in abundance in summer in the form of insects. If confined in close quarters, they must also have access to mineral food, such as oyster shells or crushed bones, with gravel and sand.

870. Of the many varieties of fowls, the dorkings, the game and the black Spanish, may be considered as among the most useful and profitable. As a market fowl, the dorking is probably unsurpassed, but the choice of the variety is generally a matter of individual fancy.

CHAPTER XXV.

THE ECONOMY OF THE FARM.

871. The success of the farmer will depend more on the general management of the farm, than on knowledge or skill in any one particular department. It is evident from the preceding pages that to make the greatest profit he must have a greater variety of knowledge, and more judgment and common sense than are required in any

merely mechanical employment, and without constant thought in planning and directing, he will constantly fail to attain the desired result, notwithstanding the most untiring industry.

872. The choice of a location should be well considered, and it is especially important whether it be near or remote from market, since the particular branch of farming to be followed will depend a good deal on market facilities. The quality of the land should be taken into view. The best lands will command the highest price, other things being equal. But it will probably be found to be better to buy good lands, though the original cost be greater, than to spend one's time and energies in tilling a poor soil simply because it is cheaper. The profit to be derived is far greater in proportion on the former, and the original cost is paid off more speedily and easily.

873. The location of the buildings requires careful consideration. How much time and strength will be wasted every year if the buildings be unnecessarily so placed as to require expensive teaming to and from the fields, or the barn and outbuildings so situated as to occasion many unnecessary steps, when a more judicious location would have avoided all? These points have a direct and important bearing on the profit to be derived from farming.

874. Then as to the fences required, both along the public ways and along division lines. What are the most economical? They should be constructed according to the purpose for which the land is to be used, whether for the general culture of farm crops, or for cattle or sheep husbandry. They can be built when other and more important labors are not pressing. But it should be remembered that all useless and unnecessary fences

involve a positive loss, as they are kept up at a constant expense, be it more or less, to say nothing of the constant loss of interest on the original cost, and the loss of the land they cover, which in many cases is no small item.

875. It is not good economy to use old and worn out or otherwise unsuitable implements on the farm, nor should shovels, hoes, ploughs and other implements requiring strength for their use, be heavier than is necessary to accomplish the object desired. Good implements save labor, while those ill-suited to the purpose increase it.

876. But though the best are, on the whole, the cheapest, even if the first cost be greater, yet it does not follow that they should be bought beyond the actual wants of the farm. Expensive implements that are rarely used, increase the permanent investment, and occasion great inconvenience, by requiring much space and care. They should not therefore be accumulated on the farm merely because they are new and good in themselves. If they are not wanted, the money paid for them is often worse than lost.

Fig. 88.—Mowing Machine, in operation.

877. Some may be needed but a few hours in the course of the year, and yet, for that time, may be of the highest importance. In such cases, where the farm is not large enough to make it necessary to own the implements, two or more neighbors can buy and own them in common. The mowing machine, (Fig. 88,) the reaper, the stump

puller, the stone lifter and the threshing machine, in a section of small farms, may be obtained in this way.

878. The storage and preservation of implements require thought and attention. Exposure to the weather will often rust and otherwise injure farming tools, while a little care will preserve them. Some system of management should be adopted for saving the more expensive ones from unnecessary injury.

879. The cost of a well-arranged tool room will not seem great, when we consider its convenience, and the saving which may be effected by it. "A place for every thing and every thing in its place," is a maxim nowhere more important than on the farm. On many farms much time is wasted in searching for tools left out of place and ill cared for, which should be saved.

880. A mistake not unfrequently made by farmers, is that of undertaking more than their capital will warrant. Profit depends more on thoroughness and quality of cultivation than on the quantity of land put under tillage. If a man has a large capital, can employ a strong force, and has the capacity and industry to direct extensive operations, he can cultivate a large farm, perhaps, to a profit. But if he has only a small capital, and is mainly dependent on his own labor, he should limit his operations accordingly.

881. This error of undertaking too much, often occasions the waste of many things, the value of which, in the aggregate, would amount to a good profit on the whole capital invested in the farm, if the waste were avoided. For want of means, the farmer is often obliged to sell at low prices, and buy at unfavorable times. This, perhaps, leads to a failure, or at least makes life uncomfortable, when the same knowledge and energies on a smaller farm would have obtained complete success.

882. After expending time and labor, both of which have a distinct money value, in ploughing and planting, none can doubt that it is good economy, after the crops are well started, to guard them carefully against their various enemies, and to give the additional time and labor necessary for this purpose.

883. After corn is up, for instance, it is worth while to protect it from birds and insects. So it is time well spent to examine every hill once in every three or four days till it is well grown, to arrest the work of the cut worm, found at the root of many a plant. If taken in season, he can do little injury. The plant will give a sure indication of his presence before it is entirely cut off and destroyed. It is important also to examine the trees of the orchard, and dig out the borer.

884. Great losses might be avoided, if a regular system like this were adopted with regard to every crop. If it is worth planting, it is surely worth the trouble of protecting.

885. The wastes of the farm are innumerable. Mention has already been made of losses arising from badly arranged and ill-constructed farm buildings, but perhaps the want of economy and skill in the management of fertilizers, is a source of greater loss than any thing else upon most farms.

886. No matter what particular course of culture may be adopted, it is only by the application of a sufficient quantity of fertilizers, of the right quality, that the farmer can keep up and increase the fertility of his land, and cause it to produce more abundant crops every year.

887. The utmost knowledge and skill should, therefore, be directed to the increase and preservation of every thing that can be turned to good account. Let nothing be wasted. Draw from the muck bed, or from any retentive

subsoil, a sufficient quantity of absorbents to mix with the materials in the barn cellar.

888. A compost may be formed of bones, ashes, old mortar, dead animal matter, loam, scrapings from the road side, and many other things worth saving, and if the run from the sink-spout and the water from the wash tubs, could be directed upon such a compost, a large amount of valuable manure might be added in the course of the year, to that now made on most farms.

889. The most direct method of increasing the fertility of the farm, is the keeping of a great number of cattle, feeding them well, and supplying a great deal of litter. With an abundance of grass, the farmer can keep more cattle; with well fed cattle he has more manure, and with this he can increase his crops. But it should be remembered that no more stock ought to be kept than can be well fed.

890. If a farm is to be stocked to its utmost capacity, green fodder should be cultivated, and it will be found advantageous to devote a considerable space to corn, to be cut up and fed green, and to clover and root crops. If the stock are kept in the barn or in small lots near at hand, the manure may be saved and increased by adding loam and other materials, while the outlands may be kept in grass and made to produce abundant crops by means of liberal top dressings.

891. The losses arising from wintering stock poorly, and from injudicious feeding in general, are vastly greater than most people suppose. Even where working and fattening cattle are well sheltered and well fed, young stock often have but little shelter, with coarse swale hay or straw to eat, and are left to take care of themselves. Young animals should be kept growing rapidly, so as to

develop their muscles and increase their size. They come to maturity earlier, and yield more profit, when well taken care of. In their case bad treatment is the worst possible economy. They must have nutritious food and enough of it, if any profit is to be derived from them.

892. Among the wastes of the farm may be mentioned the spaces along division walls, so often grown up with bushes and entirely lost to cultivation, giving an unsightly appearance to the lot, and forming a seed-bed for weeds. Many a load of rich loam might be taken from these head-lands and spread upon the rest of the piece, to great advantage.

893. Some farmers make a practice of throwing the small stones on the stubble lands into heaps upon the grass, and letting them lie there to be mown over year after year. In many cases they are not removed till the land is ploughed up again. No man who manages in this slovenly way deserves to succeed.

894. A garden should be found on every farm at a convenient distance from the house. This is too often neglected, though it pays a greater profit, if its produce be estimated at its fair market value, than any other portion of the farm. An abundance of vegetables, of various kinds, both early and late, does much to keep down the expenses of the table, and tends to promote the health of the family. It costs little time, and that little in the form of odd moments.

895. A hot-bed is a convenient means of starting many early vegetables, either for market or for family use. It may be made at a season of leisure, and costs but little.

896. The loam to be used for this purpose, should be selected and thrown into a heap in September. The construction of a frame may be deferred till winter.

897. To make the frame, take two-inch stuff and spike it to corner posts or joists, making the back side twice as high as the front, so as to give the proper inclination to the sashes. The frame may be four or five feet wide, and nine or twelve feet long. If the back and front are fastened by iron bolts and screws, the frame can easily be taken to pieces and laid away when not in use.

898. A bed of nine feet long will require three sashes. Where the sashes meet, a piece of wood three inches wide and two thick, should be set in from the back to front for the sashes to run upon, and it may extend back a foot or two beyond the body of the frame.

899. Select a south-east exposure. Dig down one foot, making the hole six inches larger every way than the frame. Drive down joists at the corners, and nail to their outsides two-inch plank, letting the top come up about to the top of the ground, the size of this structure corresponding to that of the frame, so that the latter will set firmly upon it. The bed itself should be made about the middle of March.

900. For the heating material, take coarse fresh manure from the horse stables, shake it up well and mix thoroughly, then put it evenly into the bed, beating it down with the fork, but not treading upon it. Raise it up two feet or so, the back part higher than the front, and make the whole about six inches higher than it is intended to have it stand, to allow for settling.

901. To get a steady and long heat, alternate layers of tan bark and manure may be used, or a mixture of leaves with the manure, will do. Something of the kind is important, to make the heat hold out well.

902. The sashes may be put on after the bed is formed, and the heat will begin to rise in two or three days, when

the sash may be slightly raised to let the steam pass off, and soon after the loam may be lightly spread over the manure to the depth of six or seven inches.

903. A day or two after the loam has been added, the bed will be ready for the seed, which is generally sown in drills across the bed.

904. Sometimes the manure ferments so rapidly as to give out an amount of hot steam sufficient to destroy the roots of tender plants. This danger can be avoided by sowing the seed in small flower pots set into the soil up to the rims, which may be raised when the heat is too intense, and lowered again as it moderates.

905. The same object may be effected by thrusting down a large stick in several places in the bed, and withdrawing it, leaving open holes which will soon lessen the intensity of the heat.

906. A sharp pointed stick thrust down into the manure and allowed to remain a few minutes, will show well enough the degree of heat there.

907. But constant watchfulness is required to secure such ventilation as will prevent over-heating and a feeble growth, and the frames should be open at proper times for this purpose, but the external air must be let in cautiously, and only when it is not very cold, or the plants will be injured by the chill.

908. Cucumbers and similar plants may be sown on pieces of inverted sod in the bed, when they are to be started early; they can then be removed to the garden without injury as soon as the season admits of it.

909. Cabbages, cauliflowers, melons, tomatoes, peppers, celery, lettuce and many other plants, may be started in the hot-bed, to be transplanted to the garden as soon as the season is far enough advanced.

910. Hot-beds heated by hot water or steam can be more easily regulated, but the plan described above is the simplest, cheapest, and often the only practicable method on the farm. Even with this simple arrangement, however, care and experience are necessary to secure success.

911. The culture of fruit is of itself sufficiently attractive to secure some attention. But too many manage their orchards as if they thought it enough to set out the trees, without bestowing any care upon them afterwards. There is no economy in buying poor or even second-rate trees. Get the best and set them out in the best manner. But one or two standard varieties known and esteemed in the market, are far more profitable than a great many.

912. Young fruit trees pay well for great care and attention. Enrich the land, therefore, and keep it under high cultivation for the first few years. After the trees have come into bearing, no exhausting crops should be allowed under them, unless manure enough is used for both. It is not well to starve fruit trees for the sake of a less valuable crop. But some of the smaller fruits like currants, raspberries or blackberries, all of which admit of partial shading, may be tolerated in apple and pear orchards.

913. If trees are found to be making wood too fast to bear fruit well in rich and highly tilled soil, laying down the land to grass is generally enough to check their too rapid growth, and bring them into a bearing condition. If the land be already in grass and a greater growth is desired, the grass may be spaded up in a circle of ten or twelve feet from the tree. The rootlets extend out in every direction as far as the ends of the branches, and often farther. A foot or two spaded up round the tree is, therefore, of very little service. But the surface soil

under fruit trees should not be stirred to a depth of more than four inches. It is better to manure on the surface.

914. Pruning should begin while the tree is young, but little being done at a time, and should be continued when necessary to bring the tree into proper shape. If a young tree is trimmed, the activity of the sap soon heals up the wound. Not so an old tree. The best time to prune fruit trees is late in the fall, or early in winter before the sap has started, or in midsummer after it has thickened so as not to flow rapidly. But pruning may be done at any time during the year except March and April, when it should be avoided both for fruit and ornamental trees.

915. Apples and pears should be taken from the tree before the ripening process has advanced far. A summer pear fully ripened on the tree, is very inferior to one ripened in a cool, dry place not exposed to the air. The natural process of ripening on the tree appears to benefit the seed merely, while woody fibre is rapidly formed in the fruit, but if the fruit be taken off and laid away just before beginning to ripen, sugar and juice are elaborated instead. Pears otherwise inferior may thus be made juicy and delicious.

916. It is easy to have a constant supply of healthful fruits through the season. The strawberry deserves more general and careful attention than it receives. After the crop has been picked in June and July, let the runners spread, and give them a deep rich soil to strike into, merely thinning out the weaker ones. In this way the vines are easily renewed from year to year.

917. The raspberry and the blackberry may stand under trees, or along the sides of walls or fences. When they have done bearing, the old fruit stalks should be cut

out and a few of the weaker canes also. Six canes of the new growth to the square foot may be allowed to stand, and perfect themselves for next year's bearing. It is well to lay them down and cover them over with straw or earth, as a winter protection.

918. The gooseberry does best in a moist situation, somewhat shaded. Dry hot weather, if exposed to the direct rays of the sun, often causes it to mildew. A heavy mulching of salt or meadow hay around the roots, is useful to it. A mulching of old hay or straw about the roots of all trees and shrubs enriches the land, and prevents the ill effects of a summer drought.

919. Grapes should be set about the twentieth of October, if convenient, but they do very well if set out in spring. The best time to prune or cut them in, is in the month of November. The first year after they are set out they may be allowed to run at random, to be cut back to within eighteen inches or two feet of the ground in November. The object is to get a strong and healthy growth of wood before they are brought to bearing freely.

920. Trees planted for ornamental purposes around the house and along the road-sides, add not only to the beauty of the homestead and the landscape, but to the real and permanent value of the estate, and thus pay well for the labor and care bestowed upon them.

921. The negligence as to cutting grass and grain at the proper season, and allowing it to get too ripe, is a source of very serious loss on many farms. The time of cutting wheat and all the other grains very materially affects the proportion of flour and bran, or the finer and coarser parts in the flour or meal. The grain is heavier, sweeter and whiter when cut ten or twelve days before

full ripeness, than if allowed to reach perfect maturity. It also measures more and makes more flour.

922. When the grain is still soft or in the milk, it contains but little woody fibre. Starch, gluten and sugar, in which the nutritive value consists, are then most abundant. As the ripening process advances, the woody fibre increases. The skin or outer covering of the grain rapidly thickens, and loses its fine color. It assumes a dull and husky appearance in the bin, if allowed to ripen fully, and is really worth considerably less than if cut at the proper season.

923. The same is true of all the small grains. Oats especially, the straw of which is fed to stock, should be cut while still green, or when only slightly turned. The early cut yield as much and as plump grain as those which get dead ripe, and the straw is far more valuable.

924. The keeping of accurate accounts is indispensable to complete success in farming. Without them the farmer can never see just where he stands, or whether he is making or losing money by this or that course of culture. It is well to keep a separate debit and credit account for each lot, charging it with all that is expended upon it from time to time in labor, manure and seed, and crediting it with the crops produced. At the end of the year the balance will show at a glance the gain or loss for the season.

925. And so let a separate account be kept for each department, a stock account, an account of household and personal expenses, &c. In this way a much better idea can be obtained of the actual state of our affairs at any particular time, than in any other.

CHAPTER XXVI.

ECONOMY OF THE HOUSEHOLD.

926. The success and profit of any farming enterprise will in many cases depend very much upon the thrifty and judicious management of matters within the house. The exercise of skill, prudence and good judgment on the part of the farmer's wife, is called for in a thousand ways.

927. Take the dairy as an example. Costly barns, well-selected cows and judicious feeding in the butter or cheese dairy are of little avail, if the products are to be depreciated in value by imperfect modes of preparing them for market, where the final judgment is to be pronounced upon them, and the price will vary according to their quality.

928. The care of milk forms so important a part of the duties of every housekeeper, and it enters so largely into many processes of cooking in every household, that its character and properties should be well understood.

929. Milk is an opaque fluid of a whitish color with a sweet and agreeable taste, and is composed chiefly of caseine or curd, which gives it its strength, and from which cheese is made; an oily substance which gives it richness, and which is separated in the form of cream and butter; a sugar of milk which gives it sweetness, and a watery substance which makes it refreshing as a beverage, and which is separated from the other constituents in cheese making, and known as whey.

930. The fatty matter in pure milk varies from two and a half to six and a half per cent., the caseous or

cheesy matter from three to ten per cent., and the serous matter or whey from eighty to ninety per cent., the proportions of these several substances varying according to the kind of animal, the food used and other circumstances.

931. Though to the naked eye it appears to be of the same character throughout, under the microscope a myriad of little round or oval globules, of unequal sizes, are seen floating in the watery matter. These globules are particles of butter enclosed in a thin film of cheesy matter. They are so minute that they filter through the finest paper.

932. Milk weighs about four per cent. more than water. Cold condenses while heat liquefies it. The elements of which it is composed, being different in character and specific gravity, undergo rapid changes when at rest. The oily or butter particles being lighter than the rest, soon begin to rise to the surface in the form of a yellowish semi-liquid cream, while the greater specific gravity of the whey carries it down.

933. The butter particles in rising to the surface, bring up with them many cheesy particles, which mechanically adhere to their external surfaces, thus giving the cream more or less of a white instead of a yellow color, as well as many watery particles which make it thinner than it would otherwise be.

934. If the globules rose up free from the adhesion of other substances, they would appear in the form of pure butter, and the process of churning would be unnecessary. The collection, or coagulation of the cheesy particles, by which the curd becomes separated from the whey, sometimes takes place so rapidly, from the effect of great heat and sudden changes in the atmosphere, that there is not

time for the butter particles to rise to the surface, and they remain mixed up with the curd.

935. When exposed to a warm atmosphere, milk readily becomes sour, its sugar of milk becoming what is called lactic acid. It is this sugar and the chemical changes to which it gives rise, that make milk susceptible of undergoing all degrees of fermentation, and of being made into a fermented and palatable but intoxicating liquor, which on distillation produces pure alcohol.

936. Milk will generally yield from ten to fifteen per cent. of its own volume of cream, the average being about twelve and a half per cent. Eight quarts of milk of average richness, will therefore give about one quart of cream. But the milk of some cows fed on rich food, will far exceed this, sometimes furnishing twenty per cent. of cream, and in very rare instances, twenty-five and twenty-six per cent. The quantity of cream to be obtained from milk is much more uniform than the quantity of butter from cream. Rich milk is lighter in weight than poor.

937. The temperature of milk as it comes from the cow is about blood heat, or ninety-eight degrees of Fahrenheit, and it should be cooled as little as possible before coming to rest in the pan. The depth of milk in the pan should be shallow, not greater than two or three inches. A moderate warmth and shallow depth facilitate the rising of the cream. The temperature of the dairy room should not vary much from fifty-eight degrees.

938. Milk is extremely sensitive to external influences, and hence the utmost cleanliness is necessary to preserve it for any length of time. The pails, strainers and pans, the milk room, and in short all the surroundings, must

be kept neat and clean, to an extent which only the best dairy women can appreciate.

939. The largest butter globules being comparatively the lightest, begin to rise first after the milk comes to rest in the pan, and form the first layer of cream, which is the best, since it is less filled with cheesy particles. The next largest rise a little more slowly, are more entangled with other substances and bring more of them to the surface. The smallest rise the most slowly of all, are loaded with caseous matter and produce inferior cream and butter. The most delicate cream, and the sweetest and most fragrant butter are obtained by skimming only a few hours after the milk is set.

940. On large dairy farms, a building is generally erected as a dairy house. This should be at a distance from low damp places, from which disagreeable exhalations may rise, and should be well-ventilated and kept constantly clean and sweet by the free use of pure water.

941. But in smaller dairies economy dictates the use of a room in the house. This should be, if possible, on the north side, and used exclusively for this purpose. Most cellars are unsuitable for setting milk, but where a large and airy room is partitioned off from the rest of the cellar, and can be thoroughly ventilated by windows, a greater uniformity of temperature can be secured there than on the floor above. Such a room may be used to advantage, but it should have a floor of gravel or loam, dry and porous, and without cement.

942. Carbonic acid, a heavy and noxious gas, is apt to infect the atmosphere near the bottom of a cellar, and a porous floor acts as an absorbent. It is evident that cream will not rise so quickly or so well when the milk pans are set on the cellar bottom. The air is less pure, and the

cream is liable to become acrid. When the object is to obtain the most cream in the shortest time, the milk should stand on shelves from four to six feet from the floor, around which a free circulation of air can be had from the windows.

943. A very convenient milk stand is represented in figure 88. It is made of light seasoned wood in an octagonal form, and will hold one hundred and seventy-six pans of the ordinary form and size. It is simple and easily constructed, economizes space, and may be adapted to a room of any size used for this or a similar purpose. If a stream of pure water be near at hand, it may be brought in under the stand by one channel and taken out by another, thus keeping up a constant circulation under the milk stand. This is regarded as highly important by many dairymen.

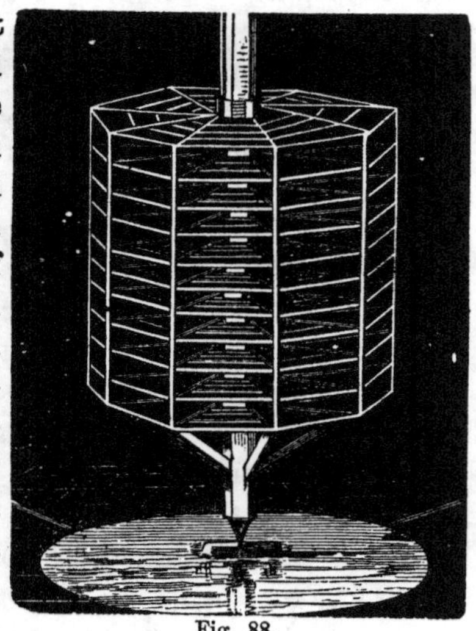

Fig. 88.

944. Milk pans are generally made of tin, this having been found to be the best on the whole. After the milk has stood from eighteen to twenty-four hours in a favorable place, the cream may be removed and placed in stone jars where it is kept till the churning. It is always best to churn as often as possible; in large dairies every day, in smaller ones every other day. But where this is not practicable, put the cream into a stone jar and sprinkle

23

over a little pure fine salt. When more cream is added, stir up the whole together and sprinkle over it a little more salt, and so on till there is enough to churn.

945. Butter may be got from cream when at a temperature ranging from forty-five to seventy-five degrees Fahrenheit, but it is a matter of the utmost nicety to regulate the temperature so as to get the best quality of butter from it. Careful experiments have seemed to show that the cream being at about fifty-one degrees at the beginning of the churning, the best quality of butter may be obtained from it. The temperature rises during the operation several degrees, depending much on the time it takes. If it were fifty-one or fifty-two degrees at the beginning, it would be about fifty-five degrees at the close. But if the object be to obtain the greatest quantity of butter from cream, the churning may be commenced with the cream at fifty-six degrees, and the temperature will gradually rise to about sixty. The greatest quantity of butter of the best quality, is got from cream standing at about fifty-three degrees. To bring the cream to a proper temperature it may be lowered into the water in a well and remain over night in hot weather, or receive the addition of a little warm water in winter.

946. The operation of churning should not be hurried. The butter from cream churned from a half to three-quarters of an hour, is of far better quality and consistency than that churned in five or ten minutes, in which time it may be brought with a higher temperature of the cream.

947. A simple square box turning on an axle is one of the best forms of the churn. It is the concussion rather than the motion which brings the butter, and this form of churn gives it as well as the dasher. The cream

takes a compound motion, and the concussion against the sides and right angled corners is very great.

948. After the butter has come, it must be thoroughly worked till the buttermilk is removed. The best way of doing this is on the butter worker, (Fig. 89.) After rolling, it may be slightly salted. A large sponge covered with a clean cloth is a most useful article for removing the milk from the surface of the butter, where it will be found to stand in little round globules after it has been pressed or worked. With the sponge nearly every particle of milk may be taken off. In warm weather have a pan of ice water at hand, and after using the sponge soak it in the water, and rinse and press it out dry to use again. Butter made in this careful way will keep better than any other, as the buttermilk, often imperfectly worked out, does more to destroy its sweetness and solidity than any thing else.

Fig. 89.

949. Another simple form of the butter worker is shown in figure 90. A plain apple tree slab is better than marble for the butter to lie on. It would not be either

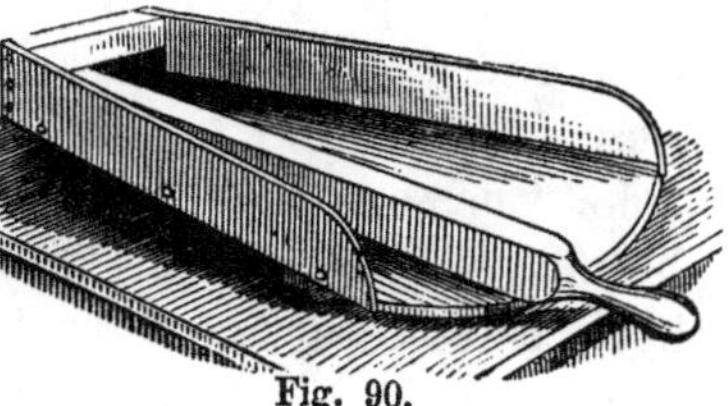

Fig. 90.

difficult or expensive to fix this upon a common table. The attachment of one end of the roller, as shown in figure 89, by a lever, is not necessary, but saves strength in working. The hands should never come in direct contact with the butter if it can be avoided, as it may be by either form of the butter worker.

950. After completely removing the buttermilk, the butter may be formed into pound lumps, or put down into firkins made of white oak, which should first be well cleansed. When thus made, it will keep a long time with little salting. Over-salted butter is not only less agreeable to the taste, but less healthy than that which is fresh and sweet. In general, much salt is needed only when butter is badly worked over, and to prevent the ill effects of neglect.

951. It is sometimes necessary to pack butter in new boxes, and the dairywoman should know how to prevent an unpleasant flavor from being imparted to the butter by the fresh wood. For this purpose use common or bi-carbonate of soda, putting about a pound into each thirty-two pound box, and pouring boiling water upon it. If the solution be allowed to stand in the box over night, the box may be safely used the next day. The adoption of this simple precaution would often prevent great losses.

952. In medium-sized dairies the nicest quality of butter might be made from cream taken off after standing in a favorable position for twelve or eighteen hours, when the skimmed milk would still make a fine quality of cheese.

953. Cheese is made from the caseine in the milk. If allowed to become sour, milk will curdle, when the whey may be separated from it. But in practice the curd is produced by the addition of an acid in the form of

rennet, which is the stomach of the young calf prepared by washing, salting, drying and preservation.

954. Cheese may be made entirely of cream, from whole or unskimmed milk with the cream of other milk added, from milk from which a part of the cream has been taken, from ordinary skim milk, from milk that has been skimmed three or four times so as to remove nearly every particle of cream, or even from buttermilk. The acid used to curdle the milk acts only on the caseine and not on the butter particles. The latter may remain imbedded in the curd as it hardens, and will increase the richness and flavor of the cheese, but they do not add at all to its firmness, which is due to the caseine alone.

955. The process of cheese making is both chemical and mechanical. The milk is heated to about ninety-five degrees, when the rennet is added, the chemical action being thus hastened, and the separation of the whey facilitated. If the rennet be strong and good, enough may be used to curd the milk in about half an hour. It is then allowed to stand for half an hour or an hour, when it is cut across in different directions, to allow the whey to work out more freely.

956. The preparation of the rennet requires great care; indeed, every process in cheese making calls for the exercise of much judgment and experience. Many fail in consequence of hurrying the pressing. The cheese is usually allowed to stand in the press only one day, though a longer time would make a much better cheese. A self-acting cheese press is shown in figure 91.

957. A very small advance in the price of dairy products from improved quality, would add very largely to the profits of many a farm. These articles are generally the last on which purchasers are disposed to economize,

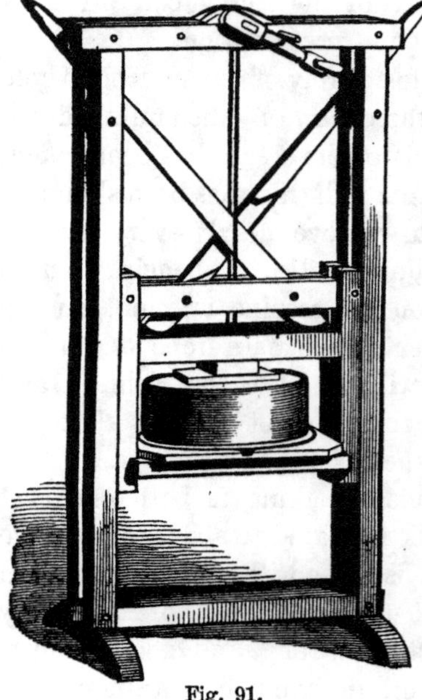
Fig. 91.

it is the quality of the articles they look at. Every thing depends on quality.*

958. There is no more important branch of domestic economy than that which relates to the use of the great staples of human food, especially the articles employed in making bread. A large part of the ill health and unhappiness of families arises from bad or defective cooking. The really good and healthy bread made in this country bears but a very small proportion to that of decidedly poor quality.

959. Undoubtedly this may in part be ascribed to the flour which the housekeeper is obliged to use. Its quality varies exceedingly in different samples, and we cannot always obtain what is really good.

960. Every hundred pounds of wheat contain from fifty-five to sixty-eight pounds of starch, from ten to twenty pounds of gluten, and from one to five pounds of fatty matter. The relative quantities of these substances vary considerably in different climates and soils. Thus the proportion of gluten is largest in wheat grown in

* The management of the dairy is stated in greater detail in the Treatise on "Milch Cows and Dairy Farming," to which any who wish to pursue the subject farther can refer.

quite warm countries. It is larger in Virginia or Maryland wheat than in that of Michigan or the Canadas.

961. Starch, as we have seen, is a white powder which forms a large part of the substance of most of the grains, as also of the potato. A general idea of the proportion in which it appears in the grains, may be obtained from figure 92, in which the grains are magnified, and where *a* represents the position and comparative quantity of the oily portions of a kernel of Indian corn, wheat and barley, the oil being in minute drops enclosed in six-sided cells, which consist chiefly of gluten; *b*, the proportion and position of the starch, and *c*, the germ or chit, which is mainly composed of gluten.

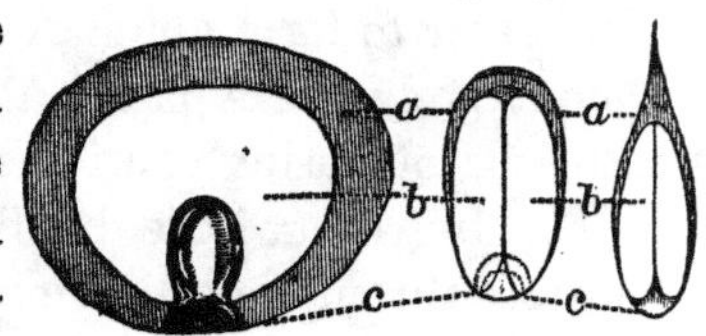

Fig. 92.

962. Gluten, as well as starch, exists in most plants, though the proportion in some is far greater than in others. It may be washed out of dough made of wheat flour, by placing it upon a sieve or a porous cloth tied over a deep dish, and pouring on water as long as it continues to run through of a whitish or milky color. The starch is carried through the cloth with the water, and the gluten is left on the cloth. The starch will soon settle to the bottom of the dish.

963. The grinding of the wheat does not wholly crush the outside covering of the grain, which is harder than the rest. This is usually sifted out from the finer portions in the form of bran, and may be fed to horses or other animals. It is often known as shorts.

964. On mixing water enough to moisten the whole mass of flour, the particles stick to each other and form

a smooth and elastic dough. This dough consists of gluten, so called from its sticky or glutinous character, and starch. These two substances, as we have seen, may be readily separated.

965. If we add a little yeast to the flour while mixing with water to form dough, and let it stand some hours in a moderately warm place, the dough begins to ferment and rise, increasing considerably in bulk.

966. In rising, little bubbles of carbonic acid gas are set free throughout the mass of dough, and this it is which makes the bread porous and light, by the stretching or expansion of the tenacious gluten. Set the dough in a hot oven, and the fermentation and rising are first hastened by the elevated temperature. But when the whole is heated up to the point of boiling water, the process is suddenly stopped, and the mass is fixed by the baking in the form it had taken when the rising was suddenly arrested by the heat.

967. But why is the rising so suddenly checked in the oven? The yeast we have added to the dough is in reality a living plant, which grows or increases with great activity when it comes in contact with the moisture of the dough, producing what we call fermentation or rising.

968. During this process, a part of the starch in the flour is changed into sugar, and this sugar into alcohol and carbonic acid gas. This gas cannot escape from the dough as the elastic gluten expands, but it remains in the shape of bubbles. At last the heat becomes great enough to destroy the yeast plant, and the process of rising ceases. The alcohol mostly escapes in the baking.

969. After the loaf is sufficiently baked, if we cut it through we find it is spongy and full of little cavities, made by the gas bubbles during the rising. It is then

soft and agreeable. But in the course of a day or two the peculiar softness disappears, and the bread seems to be drier and crumbles readily. This apparent dryness is not caused by a loss of water. Stale bread contains very nearly the same amount of water as that newly baked. Both contain on an average from thirty-five to forty-five pounds of water in every hundred pounds. Stale bread, though not generally so agreeable to the taste, is very properly regarded as more wholesome than new.

970. The more gluten any variety of flour contains, the more water will it hold. When wet, the gluten does not dry up readily, but forms a close and tenacious coating around the little cells formed in rising, which neither allows the gas enclosed in them to escape nor the water to dry up and pass off in vapor, but both are retained.

971. Now we see why flour made of wheat grown in a warmer climate and containing a larger per cent. of gluten, is sold at a higher price in the market. It is intrinsically more valuable. The larger amount of gluten not only increases its nutritive value, but its economic value also. It has a greater power of holding the carbonic acid gas produced in the fermentation, and this gives it the spongy lightness always characteristic of good bread. It also absorbs more water, and its weight is greater.

972. In an experiment said to have been carefully and accurately made, with two pounds of Cincinnati and two pounds of Alabama flour, each being mixed with a quarter of a pound of yeast, made into a loaf, and both baked in the same oven, the loaf made from the first was found to weigh three pounds, that from the second three and a half. The difference was thus about fifteen per cent. in favor of the southern or more glutinous flour. If the same

proportion were found to hold generally, six barrels of southern flour would be about equal to seven of northern.

973. Flour in its natural state contains from twelve to sixteen per cent. of water, but it will take up about half its own weight of water in addition, so that a hundred pounds of good flour make about a hundred and fifty pounds of bread.

974. It is an important fact, that the bran which is generally so carefully sifted out of the flour, is rather more nutritious than the fine flour itself. The oily parts of the grain lie mostly near the surface. The less finely bolted flour is undoubtedly more nutritious and wholesome than the finest and whitest samples.

975. Rye flour, though it does not differ materially from wheat flour in composition, is yet unlike it in some respects. Its color is not white, but a grayish brown; the bread made of it is not so porous as that made of wheat flour, nor the dough so tough. Its starch cannot be washed out like that of wheat flour. Rye bread may be kept fresh and moist much longer than wheat, perhaps on account of the peculiarity of its gluten.

976. The preference of wheat to rye arises from taste or prejudice merely. They have nearly the same nutritive value. Barley also contains about the same proportion of nutritive matter. Rye flour when mixed with an equal quantity of Indian meal, will make a very palatable and healthy bread.

977. The general principles of bread making apply alike to all kinds of flour or meal, but Indian meal, though in composition and nutritive properties not differing much from wheat flour, does not make equally spongy bread.

978. The most common modes of cooking the meats we set upon the table, are simple boiling, roasting and

baking. Out of every four pounds, beef loses one in boiling, one pound and three ounces in roasting, and one pound and five ounces in baking. The same weight of mutton loses in boiling fourteen ounces, in roasting one pound and four ounces, and in baking one pound and six ounces.

979. Fresh lean beef contains about seventy-eight per cent. of water, including the blood. Wheat flour bread, as we have seen, contains only forty-five per cent. of water. But the gluten of wheat has its corresponding element in beef in the fibrin, as it is called, and beef contains nineteen per cent. of this, while wheat flour bread has only six per cent. of gluten. Again, beef contains more or less fat, generally over three per cent. in lean beef, while we found but about one per cent. of it in the flour. The chief difference is, then, in the starch, which is not found in beef, while in bread it forms more than forty-eight per cent., or about one-half of the whole.

980. What is the fibrin of the meat? A thin piece of lean beef may be washed in clean water until its color is entirely lost, the blood being washed out, and only a white mass of fibres being left, which constitutes the muscle of the living animal. This is called fibrin. It takes its name from its fibrous nature. It contains in mixture part of the fat of the animal, and with it constitutes the main substance of the meat. Meat is therefore composed of water colored by the blood, fibrin and fat. In highly fed animals, we find the fat often collected by itself in various parts of the body, as in the suet in and around the bones, or it is deposited in large masses under the skin, instead of being evenly distributed through the fibrous mass of muscular tissue, so as to produce, in the case of beef, what is called well marbled beef.

981. The loss in cooking meat is mainly in the evaporation of water, and in the fat which melts out in roasting and baking. But this water mixed as it is with the blood, and holding more or less of various saline substances in solution, constitutes what is called the juice of the meat, and if this were all extracted the meat would become a mere tasteless mass.

982. It is very important, therefore, in cooking meats, to preserve their rich juices as much as possible. This is done in boiling and some other modes of cooking, by subjecting them to great heat when first put over the fire. By this means the fibres near the surface are contracted, the escape of the juice is prevented, and the piece is to a great extent, cooked in its own moisture.

983. Hence, if meats are to be boiled, they are usually put at once into boiling water; if to be roasted, they are exposed to a quick fire at once, either of which retains the liquid contents within, in the manner explained. If exposed to a slow fire or to cold, or only warm water, very much of the richness of meat, as well as of its nutritive quality, is lost, and the piece will become hard and dry.

984. But in the preparation of soups, broths, beef tea, &c., the object is to extract the juices; hence they are put into cold water and either simmered over a slow fire, or gradually but quickly brought to a boil. For these purposes soft water is best, because it has a greater solvent power than hard, which holds in solution more or less mineral matters, especially lime. In ordinary boiling, however, where we only wish to cook the meat, and not extract the juices in which its flavor and richness consist, hard water is better.

985. The use and manufacture of soap also form an important part of domestic economy. When oily or fatty

substances come in contact with an alkali, in solution at an elevated temperature, they undergo an entire change, and on this change the whole process of soap making depends.

986. The soap made in the farm-house is that known as soft soap, and is formed by the union of potash with more or less fatty matter. Hard soaps are made by the use of soda, with which potash is sometimes mixed. Potash will not harden when water is present, as it always is in considerable quantities in soft soap. But soap made with soda will absorb more than its own weight of water without losing its consistency.

987. The soft soaps are generally made of soft fats, while the hard soaps are more frequently made from tallow. In making castile soap, olive oil and soda are used, and its peculiar marbled appearance is produced by the mixture of iron rust. Rosin is very often added in the manufacture of common or yellow soaps.

988. Rosin soaps dissolve or form lather so readily, that they are generally believed to be very effective, but they are by no means so economical as the soda soaps, their cleansing properties being inferior.

989. The cleansing properties of soap depend mainly on its alkaline ingredients. When brought in contact with the impurities of clothing, or of the skin, which are made up of a greater or less quantity of oily matter derived from the exhalations of the body, together with dust and other foreign substances, the alkali of the soap readily seizes hold of the oily matters and dissolves or removes them.

990. If water is used without soap, it often fails to cleanse thoroughly, as it has no affinity for oily substances, and therefore leaves them and whatever has

adhered to them, in the cloth or on the skin. An alkali might be used alone, but it would be so powerful as to injure or destroy whatever it came in contact with. Washing fluids are simple solutions of caustic alkali.

991. In the life of the farmer, as in that of every other man, it is of the utmost importance to make home attractive to all the family. It is unnecessary to say that the strictest neatness and good order in all domestic arrangements, is more conducive than any thing else to this end. Without them no dwelling can have an air of cheerfulness and comfort.

992. The cultivation of flowers in the house and the garden, is well calculated to aid the skilful housekeeper in adorning and beautifying home, while it affords a pleasant occupation for leisure hours. Who does not feel the influence of flowers blooming in the window, and in the neat beds of the garden or the front yard. Graceful vines trailing over the door-way, give a charm to the poorest dwelling, and make the humblest cottage attractive.

993. The judicious, thrifty and economical management of even the smallest household, is worthy of the highest praise that man can bestow, and duties well performed, whatever they may be, give the greatest of all consolations, an approving conscience and a cheerful heart!

QUESTIONS.

[Note to the Teacher.—The questions here given are not intended for skilful and experienced teachers. They are helps, to make up for the want of skill and experience. They are to be used cautiously. They do not contain all the points in regard to which the pupil should make thoughtful inquiry and be ready to give answers; and there is always danger that, by the use of them, important things will be omitted, and mere verbal answers be given, instead of intelligent answers. Often, a single question will lead a faithful pupil to give, in his own language, the substance of a paragraph. In such a case, the particular questions may be addressed to other pupils.

For the sake of conciseness, the beginning of the questions, such as, What is? or What are?, What does? or What do?, How does? or How do?, and How is? or How are?—are often omitted, as unnecessary.]

Chapter I.—1. Agriculture? It include?—2. Object?—3. With this object, what must the husbandman have? Capital?—4. What should a complete farm have? What would be desirable?—5. Why capital?—6. Indispensably necessary to carry on a farm well? What will the farmer find by study? By practice?—7. Science? Why should a farmer have it? Whence comes science? [*Note.*—This is not all that is necessary to success. The effects of the vital action of plants, spoken of hereafter, are also to be known and considered].—8. Use of scientific knowledge?—9. Practice? Knowledge of scientific principles?—10. Why should a farmer have a good education?—11. What evidence?—12. What are the means by which the best modern improvements may be introduced upon American farms?—13. What evidence, nearer home, of the value of knowledge?—14. Advantages of a farmer's occupation?—15. Why do men of science make mistakes? What alone makes a perfect farmer?—16. Are the necessary scientific principles difficult to be understood?—17. What will be the effect of learning these principles well? Give an illustration.—18. Chemistry? A chemist? What are the objects of Chapter I.? How many objects? What does it show to be very important for a farmer? What are the two sources of the knowledge he needs? What does it show to be a noble pursuit? State what you think most worthy to be thought of and remembered.

[It is an excellent practice to call for a complete analysis of a chapter, when the pupil is capable of giving it; or to call upon one to give the several heads, and others to give an account of what is included under these heads. It is well, whenever it is possible, to get up a conversation upon the subject of the lesson. It takes time, but it cultivates the power of expression, and the valuable art of conversation; it excites an interest in the subject, and it gives the teacher opportunities to correct faults in the language of his pupils. The time it takes is not lost, but saved to the best purposes.]

CHAPTER II.—19. The air? The wind? Give an example of what is meant by elastic. Combustion? What is it necessary to?—20. Air composed of? What else? What does oxygen mean? Nitrogen? Azote? Ammonia? [Art. 32, &c. See Index.] Sulphuretted hydrogen? [Art. 40. See Index.]—21. Oxygen? An element? How abundant is oxygen?—22. Whence does oxygen come? Its tendency? Its attraction? Meant by iron's rusting? Oxide of iron? What has happened to the iron? Where is oxide of iron found? Oxides?—23. Why is oxygen called producer of acids? Acid?—24. Sulphurous acid? Sulphuric acid?—25. Water composed of?—26. Hydrogen? How heavy is it? Meant by elastic?—27. How heavy is oxygen?—28. Common air?—29. Nitrogen? Is it poisonous? What properties has it?—30. Is nitrogen always inert?—31. Protoxide of nitrogen? Deutoxide? Tritoxide? Peroxide? Nitric acid? Wonderful about this? How does this happen? The law of definite proportions? A law of nature? Can it be accidental? How general is this law? The combining number for hydrogen? Carbon? Oxygen? Nitrogen? Sulphur? Iron? Nine pounds of water made of? How many pounds of iron combine with eight pounds of oxygen? Give an example. Meant by decomposed? What has been decomposed? How much iron has been turned into rust? Atoms? How do atoms combine? How are the elementary substances represented? What is HO? NH^3? CO^2? NO? NO^2? NO^3? NO^4? NO^5?—32. Properties of nitric acid? Where is it sometimes formed?—33. Carbonic acid?—34. Wood made of? Combustion?—35. Flame?—36. Whence do the light and heat come?—37. Ammonia?—38. Where is it formed?—39. Where do roots get it?—40. Sulphuretted hydrogen?—41. What is *pure* air composed of? How much watery vapor does *common* air contain? How much carbonic acid?—42. What happens in breathing? What, in place of oxygen, is breathed out? Respiration? Is the quantity of air rendered unfit for respiration known? How much pure air does a man need? How many cubic feet does a room, ten feet in each dimension, contain? How soon will the air in a close room of that size be rendered unfit to breathe?—43. What is meant by ventilated? How important is ventilation?—44. Do plants breathe? Is air necessary to them?—45. By daylight what do growing plants do? What is done with it? How do plants purify the atmosphere? What relation do we see in this? State it distinctly.—46. What action do plants have in the night? When is wood formed?—47. An oxide? [See 22.] Bases? What do they do? Why are the compounds of acids and bases called salts?—48. What does carbonic acid and lime form? Sulphuric acid and lime? Nitric acid and potash?—49. Oxygen combining with? Decay of fallen leaves? What are favorable? How is humus formed? Nearly all decay? What are formed during decay?—50. Humus? Ulmin? Ulmic acid? Humin? Humic acid?—51. Geic acid? Crenic acid? Apocrenic acid?—52. Action of nitrogen?

CHAPTER III.—53. What is meant by atmosphere? How high?—54. The atmosphere? What are acting in it?—55. What operations are going on in it? What are rising into it? What are all striving to do?—56. What is the sun doing? Every star? Oxygen? Water?—57. By what force does water penetrate? Capillary attraction?—58. Osmotic action? How?—59. How does oxygen act?—60. Each gas do? Give proof.—61. How does heat act? Conduction? Radiation? What effect does heat produce in solids? In liquids? In gases?—63. Attraction of gravitation? Attraction of cohesion? Force of adhesion?—64. Force of vegetable life? Force of animal life? A third force? A fourth?—65. Under what influence is woody fibre formed? What other effects has the sun's light?—66. What facts prove the influence of the sun's light?—67. What experiment shows that light gives wood its strength? *—68. Why is this? What is the difference in the growth by day and by night?—69. What power has the sun's light upon animals?—70. How important is sunshine to human beings? Point out the difference between two children, otherwise like each other, one kept much in the sunshine, the other much in the shade.—71. How does the sun act upon the soil? What precaution is desirable when trees are to be planted?—72. What other effects are produced by the atmosphere?—73. What appearance does amber or wax exhibit when rubbed? What is the cause? How is glass excited?—74. Explain what the opposite electricities are?—75. Meant by discharged?

* This is also strikingly shown in the growth of forests. Trees standing near an opening in the woods make the greatest growth of foliage and limbs, on the side towards the light.

—76. What takes place when vapor is formed? How is rain supposed to be brought on?

Chapter IV.—77. What instruments measure the changes which take place in the atmosphere?—78. On what principle is the thermometer constructed?—79. By what experiment is this proved? But what other?—80. Describe Fahrenheit's thermometer. What is meant by graduated? What is the freezing point? Go on. The boiling point? Degrees? How far need a thermometer be graduated?—81. On what principle is a barometer constructed? How can it be weighed? How heavy is air? Water?—82. The effects of the weight of the air? The purpose of a barometer?—83 How is a barometer constructed?—84. Why does the mercury rise in the barometer? How are changes in the weather foreshown?—85. What is the pressure of the air on a square inch?—86. How is fair weather foreshown? How foul weather? How a violent wind?—87. A Lowell barometer? Describe it. A vernier?—88. How far may change of weather be predicted? What has been done to discover the laws of storms?—89. Signs of rain?—90. The principle of the hygrometer?—91. How is a hygrometer made?—92. A still more delicate one?—93. The use of these three instruments?—94. What do variations in the temperature of the air depend on?—95. What changes take place in the column of air above us?—96. What do variations in the moisture depend on? Which winds are moist winds? Which dry and cold? On what else does the moisture depend? How does heat act?—97. Other atmospheric phenomena?—98. On what does the formation of dew depend? Radiation of heat? What kind of surface radiates most abundantly? Why? What happens when the sun sets? What becomes of the heat? How is dew formed? Why is it not formed in a cloudy night?—99. How is hoar-frost formed?—100. Climate of a country?—101. New England climate?—102. Influence of climate? Acclimatize?—103. Causes of diversity of climate? General? Second? Third? Another?—104. Some of the local causes? After a rain in summer, if the clouds disperse at night, why is it commonly cool? If it continues cloudy at night, why does the heat commonly continue?

Chapter V.—105. Water? How abundant is it? How important to plants?—106. A still more powerful solvent?—107. Three forms of water? On what do they depend? First form? Describe the change which takes place from the action of heat? The freezing point? Latent heat? Second form? Describe the effect of heat upon it. What is meant by *evaporate?* How much heat is required to boil water away? Third form? Latent heat of vapor?—108. Boiling? What fact is it important for a cook to know?—109. When is vapor formed? What does it, when expanding, always use up?—110. What happens when vapor turns to water? When gases are condensed? When water freezes? How can you guard vegetables from freezing?—111. Vapor in the air depend on? What happens when the air cools?—112. Clouds? Fogs? Mist?—113. The quantity of moisture in the air depend on? When the wind blows from the sea upon the land, in what case does it not rain? In what case does it rain? How is rain formed? In what other way?—114. Cause of rain in a thunder storm?—115. How is snow formed? Uses of snow?—116. How is hail formed? Hail?—117. How are springs formed?—118. Rivulets, brooks, rivers?—119. How is water important?—120. How, as a solvent? How much ammonia can it dissolve? How much carbonic acid? What else does it bring down?—121. Effect of evaporation? Of condensation?—122. Why do plants need much water? What becomes of it? Describe the experiment which proves exhalation from the leaves. What should be an object of the farmer?—123. Irrigation? Of what use is it?—to cover the hills with trees?—124. What will guard against the effects of drought? What makes soil retentive of moisture?—125. Why should the rain be allowed to penetrate the soil? What harm does it do if allowed to run off? Why should the soil be kept mellow? How should a hill-side be ploughed?—126. What are the remedies for excessive wet?—127. Drainage? How is it effected?—128. Explain the effects of drainage.—129. What others? The first? The second? The third?—130. Enumerate several important effects of thorough drainage.—131. How is deep drainage a resource against drought?—132. The best preventive?

Chapter VI.—133. How do plants resemble animals? How differ from them?—134. The simplest plant? These nourished? New plants formed? The most perfect plants increase?—135. How numerous are the simplest plants?—136. Plants next in simplicity of structure? Others? Still others? Lichens? (pronounced lyekens). 137. Most plants formed?—138. Organs?—139. The principal organs?—140. The root? It usually divide? Amount of food depend on?—141. The stem? The collar?—142. The bark?—143. The leaves? Opposite actions through their surface? The sap become?—144. The flower?—145. How do you learn the several parts of a flower? The calyx? Sépals?—146. The pétals? The corolla?—147. The stamens? The anther? Pollen?—148. Pistils? Style? Stigma? Receptacle? The style of a rose?—149. Use of pollen? Ovary? Ovules? Meant by fertilize? Embryo?—150. What commonly happens when the seeds are fertilized? Ovary become? Meant by germinate?—151. The fruit? Organs? Instances? Organic substances? Inorganic?—152. What happens when the seed is put into the ground?—153. Cotyledons? What are they unlike?—154. Dicotyledonous plants?—155. The Plumule? How does it grow?—156. Monocotyledonous plants? Describe the growth of one.—157. The organs of the stem of a tree? External? The trunk? Branches? Limbs? Branchlet? Spray? Twigs?—158. A shoot grow? Terminal bud? Axillary bud? The axil?—159. The internal organs?—160. The usual course of plants? Annual? Biennial? Perennial?—161. Why should grain be ever cut before the seed is quite ripe?—162. The necessity of classification? —163. The artificial system?—164. How are plants now divided? A natural family? Give an example.—165. A genus (plural, genera)? A species? Give an example showing the meaning of class, family, genus and species.—166. How does the practical use of classification appear? It will be well for the teacher to consider whether the names of the families ought to be learnt. If so, let them be learnt thoroughly. —167. Some of the plants which belong to the Pulse Family? Use? —168. Cress Family, character, &c., of all the rest?—169. Flax Family?—170. Rose? What kind of flowers? Fruits?—171. Gourd?—172. Currant?—173. Parsley? Sunflower? Sage or Mint? Convolvulus? Night Shade? Character? Give an instance. Olive? Heath? Goose-foot? Buckwheat? Walnut? Birch? Willow? Pine? Class of those that have been mentioned? Of those to be mentioned? Lily Family? Amaryllis? Iris? Orchis? Rush? Sedge?—192. Cereal grains? Character of the Grass Family?—193. Mosses?—194. Lichens?—195. Difference between a tree, a shrub and an undershrub? Undershrubs? Preparation for planting of perennial plants?—197. Alimentary plants? Forage?—198. Origin of cultivated plants? Of Indian corn?—199. How have they been improved? Give examples. The most striking?—200. Whence come the good qualities of most cultivated plants? Of animals subject to man? What is likely to happen to children left to themselves?

CHAPTER VII.—201. Essential to the formation of plant-cells? What come from carbonic acid?—from ammonia? The history of CO^2 and NH^3? Whence may O and H come? N and O?—202. What power must the simplest plant have?—203. The osmotic power? Experiment shows its action? Endosmose? Exosmose? Effects are produced by this power in plants?—in animals?—204. Why must water be abundantly supplied to growing plants?—205. The most indispensable article in the food of plants? What does it do? Of what are the solid parts of plants formed?—206. Describe ammonia. What is it? How essential is it to plants?—207. Atmospheric food of plants? The four essential elements?—208. Is it possible for plants to grow without any connection with the earth?—209. Is the C in plants pure? Charring? What is done by charring?—210. Peat? Anthracite and bituminous coal? What may be seen in them?—211. What happens in charring? What is consumed? What happens when charcoal is burnt in air?—212. Combustion? The combustible or atmospheric elements? The incombustible or mineral elements? How many are found in the ashes of every plant? The ashes of plants of particular families remarkable for?—213. The ashes of plants of the Cruciferous Family contain? Cruciferous Family?—214. The ashes of plants of the Pulse Family? Sulphate of lime? Lime?—215. What is found in the ashes of wheat and other grain? Phosphate of lime? Phosphoric acid? Phosphorus? What tendency has it to combine with O? What use is made of it? Why is phosphate of lime called bone earth?—216. In what plants is silica found? What properties does it give? Silica? Describe it. Why is it considered an acid? [See, also, art. 270.]—217. Where is potash found? How is it obtained? Soft soap formed? Pearlash?—218. Common potash? Potassa? Potassium?—219. Soda? Where is it found? Sodium?—220. For what have the ashes of sea-plants been valued? Whence comes the name alkali? The alkalies? Their properties?—221. How are they acted on by acids? Give an example. Why is the salt formed called a neutral salt? In what state do most mineral substances enter plants?—222. Some plants growing near them contain? Common salt? Chlorine? Very remarkable?—224. What other oxides are found in the ashes of plants?—225. Magnesia?—226. The ashes of plants growing in the sea or on the sea-shore contain?—227. Iodine? Why is it used in the processes of the daguerreotype?—228. In what state does iodine occur?—229. Bromine?—230. What other metals are found in plants?—231. The acids most important in the structure of plants? What other acid is essential to some plants?—233. Are these the only acids found in plants? Malic acid? Citric acid? Oxalic acid? Tartaric acid?—234. In what form are they found in plants? Potash plants? Why are they so called?—235. Silica plants? Why so called? Lime plants? Why?—236. What phosphates are found in all plants? In what particularly? What is a phosphate?—237. Why are the atmospheric elements so called? Why the earthy?—238. What else are found in plants? Fluorine? Fluoride of calcium? Mineral food of plants?

CHAPTER VIII.—239. How large are the particles of the elements of plants? Organic compounds?—240. What is meant by organized?—241. The most important organic compounds?—242. Cellulose? Why is it called woody fibre?—243. Vegetable Jelly?—244. Starch? Where is it found? How important is it as an article of food? Whence is it obtained for use in the arts? What use is made of it?—245. Gum? Dextrine?—246. Sugar? Sugar candy? Muscovado sugar? What plants yield it most abundantly?—247. Vegetable oil? From what is it extracted? Linseed oil? Colza?—248. Wax? Beeswax?—249. What relation exists between these compounds? Effect in the animal system?—250. Nitrogenous Compounds? Why so called? What are they in their simplest form? Why called protein?—251. Protein form? Effect in the bodies of animals?—252. Casein?—253. Albumen?—254. Gluten? What plants contain it abundantly?—255. Importance of this fact?—256. Whence must every thing in the body of an animal have come? Whence, previously?—257. Chlorophyl?—258. Whence the yellow colors? The rich autumnal colors?—259. Whence coloring substances?—260. Tannin? Properties?—261. What powers seem to be acting in plants? How?

CHAPTER IX.—262. What is known of the interior of the earth? The crust? An element? How many have been found?—263. How many of them are metals? Some of the more common metals? Describe such as you know. In what state are most of the metals found? Ores?—264. Found pure? Aerolites?—265. Geology? A geologist? Chemistry? A chemist? Chemical analysis?—266. The soil? The three most important earths?—267. From what has the soil been formed? The principal rocks?—268. Granite? Sienite? Greenstone? Trap-rock? Gneiss? (pronounced *nice.*) Mica Slate? Porphyry? The Granitic rocks?—269. Their qualities? Composed of?—270. Silex or Silica? Why is it called silicic acid? When pure, called? How abundant is it? Beautiful stones are silica? How, in the processes of nature, is it dissolved?—271. Its usual form?—272. Silicious Rocks? Soil is formed of these two kinds of sand?—273. Slaty or Argillaceous Rocks? Clay? What kind of earth does it form? Colors? Properties?—274. Kinds of clay? Made of clay? Plastic arts? Kaolin? Petuntze? Uses?—275. Aluminum? Alumina? The sapphire and the ruby? The topaz? The diamond?—276. Calcareous Rocks? Chalk? Sort of soil is a calcareous soil?—277. Pudding-stone Rocks? What holds the materials together? Soil is formed of them?—278. What has happened to all these rocks? What six substances does the sand formed from granitic rocks contain?

Whence clay? Lime?—279. Weathering? How does heat act?—rain?—frost? What else are acting? What action has water?—280. The best natural soil? Most fertile soils?—281. Basis of all soils? A clayey sand? A sandy clay? Loamy clay?—282. What must be added to all these to form a good soil? How does humus act? Humus? What is land which contains four per cent. of it?—eight?—ten?—283. How may it be found whether humus is present in a soil?—284. When, only, is humus unfavorable to useful vegetation? How must they be corrected?—285. How does humus act? What are successively produced?—286. What action has the carbonic acid? The oxygen?—287. What other important property ought a soil to have? What will give this property?—288. The richest natural soils? Where are such soils found? Describe the formation of these soils? In what parts are the most fertile of such soils found?—289. The soils next in value? Why?—290. Of what are granite, gneiss, and mica slate, composed? Syenite? Trap-rock? Greenstone? Porphyry? Examine the table and give it on the black-board.—291. Quartz? Felspar? Mica? Hornblende? Augite?—292. What is to be done when clay can easily be had? An amendment?—293. A more effectual way of rendering sandy soil fertile? Where is this to be found? How does it act? Describe it. What will be the effect of this process repeated?—294. What substitute can be found for clay and marsh mud?—295. What do these amendments require?—296. How is a clayey soil to be improved? When may this be done? Effect of the sand? How else?—297. How may a calcareous soil be amended? A fourth kind of soil? They really? How are they to be managed?—299. What manures are good for such a soil?—300. Heavy, cold soils? Why are wet lands cold?—301. Effect has color? Dr. Franklin's experiment?—302. Is our soil formed by the crumbling of rocks just beneath the surface? Diluvial soils? When the forests are cut down what kind of soil is usually found? How fertile? How has this soil been formed?—304. Give instances of the effect of weathering. Of oxygen; carbonic acid; in water. How powerful is its action?—306. How else does CO^2 act? How do the alkaline carbonates act?—308. The remedy for the loss of fertility?—309. Is a soil barren for one plant also barren for every other?

Chapter X.—310. Subsoil?—311. Its influence? What is to be done? Why should it be done gradually?—312. When should the subsoil be mixed with the soil? Evils of a subsoil impermeable to water? The remedy?—313. Indications of this evil? Remedy? Rule?

Chapter XI.—314. Double part does a soil play? Qualities do different plants require?—315. How are these evils remedied? Amendments? Give examples.—316. Amendments for argillaceous soils?—317. For silicious soils? Irrigation? Its effects?—318. Effects of planting? Draining?—319. As to the advisableness of an amendment, what question is to be settled?—320. A permanent amendment? One not necessarily permanent?

Chapter XII.—321. What elements must a soil contain? For marine plants, what are necessary?—322. In what state are the elements found? A silicate? Sulphate? Nitrate? Phosphate? Carbonate? A sulphate of lime, &c.? How are these known to be essential to plants? The effect if one were entirely wanting? Give an example.—324. Why should the elements derived from the atmosphere be artificially furnished?—326. What was often said by Prof. Nuttall?—327. Describe the slow process by which wild plants are furnished with humus.—328. How is forest mould formed? When it is quite wanting how is humus to be furnished? Organic substances?—329. How do they act?—330. How do they act on each other?—331. Fertilizers divided?—332. What double character have mineral fertilizers?—333. The principal m. f.?—334. Quicklime? Its effects? What state should it be in? Power has water containing $C O^2$? Act upon plants?—335. Describe a Flemish compost.—336. Valuable properties in lime?—337. Its effects in small quantities?—338. To be done with shells?—339. Marl? Its value?—340. Plaster? Why is it valuable? Sulphate of lime?—341. Where is sulphur found? Sulphuric acid? What salts does it form?—342. Hard water? How made soft water?—343. Write upon the black-board and explain the diagram showing the mutual action of sulphate of lime and carbonate of ammonia.—344. Whence comes the carb. of Am.?—345. How should plaster be applied?—346. Effects of sulphate?—347. The Westphalian proverb about ashes? Ashes? Character of these salts?—348. Unleached ashes good for?—349. Leached ashes?—350. Value of coal ashes?—351. Ley? (This word, in common dictionaries, is spelled *lie*,—in books of chemistry and agriculture, *ley* and *lye*.) Its effects? What else?—352. Value of soot?—353. Of carbonates of potash and of soda?—354. Of salts of Am.?—355. Of nitrates?—356. Phosphate of lime? To what is it essential? Value of phosphates?—357. How are these valuable?—358. The uses of common salt?—359. The object of manures? Of organic manures? Of humus? The last result of decay of vegetables?—360. In what state are organic manures to be employed? Why? Nature of Am.? How can it be saved?—361.

Organic manures divided? The principal vegetable?—362. Green manures? What plants are best for the purpose?—363. For sandy soils?—clayey?—calcareous?—364. For what districts?—365. Green crops for winter grain to be treated?—366. How weedy land to be?—367. Farther advantages of green manures?—368. Value of sea manures? Value of eel-grass? Management of sea manures? For what crops valuable?—369. What best for each vegetable? Effects of straw? Stubble to be managed? Good for hay land?—370. Said of leaves? Of different kinds?—371. Why are animal manures valuable? How do they act?—372. Flesh contain? How to be applied?—373. Best to do with a carcass? How valuable?—374. Sulphuretted hydrogen? How does it act?—375. Rank is to be given to hair, &c.? Why?—376. Effect of hair? State the loss from its being thrown away.—377. Value of blood as manure? How to be managed?—378. Of bones? How prepared and applied?—379. Action be accelerated?—380. Whence the value of animal manures?—381. Best of manures? Why? Value of milk? Different kinds?—382. Why is particular care to be taken of manure? To be saved? The dangers of loss?—383. Loss be prevented?—384. Decomposition? Cause? Vital principle act?—385. Brings on vinous fermentation? Its products? How is vinegar made?—386. Final products of decay? Essential to it? Prevents it?—387. The stable to be kept sweet? Cost?—388. Different value of different manures? Should all be thrown together?—390. Liquid manure formed? For what fields to be used?—391. Best of all?—392. Says chemical analysis? How lost? Consequences?—393-394. Prevent disagreeable effects?—395. Guano? Where found? Consist of?—396. Experiment. Another.—397. Another. Common fear? Well founded?—398. Where may plants get their mineral elements? Evidence?—399. Those eight? Still more essential?—400-401. How valuable is muck? Substitutes? From ponds and pools? How save the scourings of hills?—402. Good compost for trees? Clayey soil? Common crop? Another? How is health affected by the management of manure? Effect of these gases on health? Give instances. Of carbonic acid?—404. In small quantities? Effects in the school room? Of ammonia?—405. Where are they generated? How show themselves? Who are particularly subject to fever and other disease?—406. How may the well be affected? Consequence? Preventive?

Chapter XIII.—407. When is draining indispensable? The other essential operations? Ploughing?—408. Objects of ploughing? How can most of these be done most effectually?—410. Advantages of deep ploughing? Effects of deep ploughing upon the roots?—411. Saving

does it produce? How? How must the change be made? On what will the number of rootlets depend? When is burning useful? What ashes contain most potash? The rule?—412. Use of thorough tillage? What seems to show this? The reason?—413. Value of tillage? The most important use? When may there be danger of overdoing?—414. Subsoiling? Effects? What may draw up the moisture in loosened earth?

Chapter XIV.—415. What is requisite for plant growth? What is understood by the mechanical condition of the soil?—416. Are soils naturally fitted for cultivation?—417. What condition of the soil is necessary?—418. What do most soils require? Why do they require this? What harm does standing water do?—419. Do all soils require the same preparation? What are the processes most necessary?—420. Where is clearing generally required?—421. To what process is the term clearing applied? When is it begun? How is it where the wood is valuable? How may cleared land be got into pasturage? To what should steep and inaccessible places be devoted?—422. What is next to be done? How may stumps and stones be removed?—423. How may the surface overrun with bushes be cleared?—424. How may wet lands be improved? The object of drainage?—425. What effect has stagnant water?—426. How does exclusion of the air affect soils? —427. How is drainage effected? Objection to open drains? Are they as useful as covered ones?—428. How are the objections to open drains avoided? How are underdrains made? The advantage of tiles?—429. How must a stone drain be laid? Is it ever good economy?—430. How is it with a tile drain?—431. Economy of the tile drain?—432. What is the pipe tile?—433. What fall is necessary? —434. What is the sole tile? The objection to it?—435. What is a brush drain? How is it laid? Is it efficient?—436. What distance apart may drains be laid? How do stiff and porous soils affect the distance of the drains?—437. What depth is required?—438. What is the effect of thorough draining? How does the air affect the soil?—439. With what is the atmosphere charged?—440. What effect has draining on the temperature of the soil?—441. What is the result of increasing the temperature?—442. How does draining affect the texture of the soil?—443. How is land got ready for planting?—444. What implements are used?—445. What is the spade?—446. The implement in most common use?—447. How does the plough operate on the soil? —448. How should the furrow be turned?—449. What about the depth? —450. What is the effect of bringing up the subsoil?—451. Advantage of deep ploughing?—452. Where is it especially necessary?—453.

What is the design of the subsoil plough?—454. Benefits of subsoil ploughing? What do recent investigations show?—455. What is the Michigan plough?—456. What is the digger?—457. How is the harrow used?—458. How does the cultivator compare with the harrow?—459. How is the roller used?—460. How is it useful in laying down land?—461. What caution is needed? Effect of rolling stiff soils?

CHAPTER XV.—462. What conditions are requisite to germination? How does light affect germination?—463. What is the first process?—464. How does the plant grow?—465. How does it draw food from the air? What office do the leaves fulfil? How numerous are the breathing pores of the leaves?—466. What do plants usually spring from? What is a tuber? What is a bulb?—467. What is a seed bed?—468. What do all plants require for their complete development? What points is it necessary to attend?—469. Can an imperfect seed grow?—470. What is the advantage of perfect seed?—471. How may good seed be known?—472. How may the germinating power of seed be ascertained?—473. What advantage is there in learning the quality of the seed?—474. How does the vitality of seeds differ?—475. How can the appearance of plants on new soil be explained?—476. How is it with seeds of the turnip and the grasses?—477. What change takes place in seeds? How can deterioration be prevented?—478. Why is change of seed necessary?—479. When may a change be avoided? What care is requisite to preserve the purity of seeds?—480. What common examples can be given of seeds producing fruits different from those which produced them?—481. How may varieties be obtained?—482. How may varieties of Indian corn be produced?—483. Is a modification of the original fixed at once?—484. How is it in cultivating potatoes?—485. How may new varieties of the potato be produced?—486. Is any new variety obtained from layers, or in grafting or budding?—487. How may we judge of the quantity of seed required?—488. How is the seed required influenced by the soil?—489. What other conditions affect the quantity of seed required?—490. How does the mode of sowing affect it?—491. How is it with thin and thick sown crops?—492. When may steeping seeds be practiced?—493. What is best adapted to nourish the germ of plants?—494. What circumstances affect the time of planting?—495. How does the condition of the soil affect the time of sowing?—496. Is there any general rule applicable to all crops?—497. To what depth should seeds be covered?—498. The depth of clay and sandy soils?—499. How does the size of seeds affect the depth of covering?—500. What are some of the modes of sowing? What of broad cast and drill sowing?—501. Advantages of drill sowing?—502. How are seeds covered?—503. How is Indian corn planted?—504. How is the manure applied?—505. What is transplanting?—506. What advantages has this mode of culture?—507. What is needed to transplant successfully?—508. At what stage of growth is transplanting to be performed?—509. What care is required in transplanting older trees?—510. How may injury to the roots be remedied?—511. How does the removal of part of the top affect the tree?—512. How may trees and shrubs be grown?

Chapter XVI.—513. How are cultivated plants divided?—514. What do the cereals include? Is buckwheat a cereal plant?—515. What is one of the most important of cereals?—516. What soils does Indian corn require?—517. How is the land prepared?—518. How is it with stiff soils?—519. What manures should be used for this crop?—520. How are they to be applied?—521. Manure having been applied in the fall, what course is adopted in spring?—522. What is the objection to putting all the manure in the hill?—523. Another objection?—524. How may concentrated manures be used to advantage?—525. How may the land be put into good condition? Why is it poor economy to raise poor crops?—526. What is of special importance?—527. What is the next step?—528. Distance of the hills for corn? Advantage of using the corn-planter?—529. Why should the plants stand closely?—530. When may soaking the seed be resorted to? What is the best substance to soak seed corn in?—531. Is drill planting recommended?—532. How deep is it covered?—533. When is the crop first hoed? What implement is used between the rows, and how?—534. What implement is used in subsequent hoeings, and why?—535. How many hoeings are necessary for corn?—536. How is the seed to be selected?—537. When are the top stalks cut? Effect of cutting too early?—538. What is the better practice?—539. The cheapest way of stooking?—540. What are the varieties of wheat?—541. How does winter differ from spring wheat?—542. What soil is adapted to wheat?—543. What is requisite in the soil?—544. What preparation of land is needed?—545. How is it sown?—546. What are the advantages of drill sowing?—547. Are there any other advantages?—548. What quantity of seed is sown per acre?—549. Where does wheat come in the rotation?—550. What is important?—551. When is it harvested?—552. What takes place if harvesting is neglected?—553. Effect of exposure to rain?—554. What is said of rye?—555. Soils adapted to rye?—556. The varieties? How does its range of culture compare with wheat?—557. Quantity of seed per acre?—558. Its use on sheep farms?—559. What of its straw?—560. What disease attacks rye?—561. Peculiarity of barley? What soils does it require?—562. When is it to be sown? —563. When is it harvested?—564. What climate is best adapted to oats?—565. How does the yield differ in different circumstances?—566. How may they be sown? Quantity of seed per acre?—567. How are oats for a green crop?—568. Effect of the roller on them?—569. When should they be cut, and how?—570. How are oats used?—571. What is said of buckwheat?—572. Where is it cultivated?—573. What soils does it do best on? How is it often used?—574. How is the land prepared for it? When is it sown? Quantity of seed per acre?—575. How is it cut and gathered?—576. How is millet used?—577. Soils best adapted to millet?—578. When is millet sown?

CHAPTER XVII.—579. What does the class of leguminous plants include? Why so called?—580. The most important of the leguminous plants? What are the varieties of the bean?—581. Soils best adapted to beans?—582. How is the land prepared? How thick should they be planted?—583. When are they to be planted?—584. When hoed and how?—585. How does the season affect the crop?—586. When is the crop to be harvested and how?—587. What yield may be expected?—588. The varieties of the pea?—589. The soil best adapted to pease?—590. What manures are used for them?—591. What cultivation is required? How are they sown?—592. How are they harvested? For what purpose is the pea often sown?—593. By what is it attacked?—594. What remedy is there?—595. What is the objection to late planting?—596. What is said of the lentil?—597. What soils does the lentil require?—598. What is said of the vetch?

CHAPTER XVIII.—599. How may the potato be raised?—600. When should seed potatoes be gathered?—601. The varieties of this plant? 602. What is a prominent constituent of the potato?—603. When is starch the most abundant?—604. What soils are best for potatoes?—605. What cultivation is required?—606. What manures?—607. What of the practice of cutting?—608. What after-culture is necessary?—609. When is the crop harvested and how?—610. What is the climate for the turnip? How do droughts affect it?—611. Advantages of its culture?—612. How should crops alternate? Effect of the root crop on the soil?—613. Varieties of the turnip?—614. Soils best adapted to it?—615. What preparation of land is necessary?—616. Effect of poor soils on the quality of the turnip? Manures best adapted to the turnip?—617. Effect of nitrogenous manures on the turnip?—618. How is the turnip sown? Objection to ridge culture?—619. Quantity of seed per acre?—620. What is the after-culture?—621. When are turnips harvested and how ?—622. How are they used? What quantity may be given to an animal ?—623. What is the kohl-rabi and how is it cultivated ?—624. Soils adapted to the cabbage ?—625. Varieties of the beet ?—626. What of the mangold ?—627. How is land prepared for beets ? How is the seed sown ?—628. What does the after-culture consist in ?—629. When is it harvested ? What care is required ?—630. What is said of the carrot ?—631. How is the carrot used ?—632. How is the carrot cultivated ?—633. What are the varieties of it and the qualities of each ?—634. The climate best suited to it ? Effect of droughts upon it ?—635. Soils best adapted to it ?—636. What cultivation is of importance ?—637. Effect of stimulating manures on it ?—638. What of the seed ?—639. Quantity of seed per acre ? Depth of covering ?—640. Preparation of land ? Time of sowing ?—641. When is it to be hoed ? Number of hoeings required ?—642. What about thinning out ?—643. When is the carrot harvested ?—544. What is said of the parsnip ?—645. How does it compare with the carrot ?—646. Varieties of the parsnip ?—647. Climate best suited to it ?—648. Soils ?—649. Preparation of the land ?—650. Its yield compared with the carrot ?—651. What of the artichoke ? Advantages of its culture ?—652. Cultivation ?—653. How is it harvested and used ?

Chapter XIX.—654. Origin of the culture of grasses ?—655. Classification of grasses ? What are the natural grasses ?—656. What are artificial grasses ?—657. Design of laying down lands with the natural grasses ?—658. What is the common practice ?—659. How do the natural grasses grow ?—660. Where is pasturage the only improvement practicable ?—661. What differences are found in the grasses ?—662. Advantages of using several species together ?—663. How is a selection made ?—664. What are some of the best grasses for mowing lands ?—665. What species are best adapted for pastures ?—666. What point is of importance in the selection for mowings ?—667. How does the time of blossoming affect the mixture for pasture grasses ?—668. Climate best adapted to bring the grasses to perfection ? How do the grasses of a dry climate or dry season compare with those of a moist one ?—669. Best time to sow grass seed ? Objection to spring sowing ?—670. Preparation of the land ?—671. How is the seed sown ? Depth of covering ?—672. Time of sowing on a clayey soil ?—673. When is clover seed sown ?—674. What are the artificial grasses ?—675. The most valuable of the artificial grasses ? Effect of clover roots on the soil ?—676. Soils best adapted to clover ?—677. What is said of the white or Dutch clover ?—678. How is lucerne affected by the climate of this country ?

CHAPTER XX.—679. How are plants used in the arts divided?—680. What is the only plant cultivated for its oil in this country?—681. In what climates does flax succeed best? The soils best suited to it?—682. How does the object of raising flax determine the soil for it?—683. What manures are good for it?—684. What preparation of the land?—685. The quantity of seed to be sown?—686. Effect of quantity of seed on the fibre? Kind of fibre most valuable?—687. How is the seed sown? What care is needed after the flax is up?—688. How is the plant gathered?—689. If allowed to ripen seed how much may be expected per acre?—690. What is said of hemp?—691. The soil best adapted to hemp?—692. How is it sown? When and how is it harvested?—693. What other plants are used in our manufacturing industries?—694. What are the varieties of osier willows?—695. What soils are adapted to the willow?—696. How is the land prepared?—697. How are the cuttings set? What attention is needed?—698. The soil best suited to broomcorn? How is the seed sown?—699. When and how is it harvested?—700. What are the varieties of the hop?—701. The soil best adapted to it?—702. What manures may be used?—703. How is the hop propagated?—704. What is said of the poles to be used?—705. When are hops gathered and how?—706. How is tobacco cultivated?—707. By what is the plant attacked?—708. When is it topped?—709. When gathered and how?

CHAPTER XXI.—710. Rotation of crops? Give an example. Object of rotation? The reason for rotation? First important principle in the rotation of crops? The second? The third? Fourth?—711. Is the order of rotation important?—712. What may be saved by rotation? How must the succession be arranged? Show, by an example on the black-board, how this may be done.—713. How may the rotation be made still longer? How may time be saved? How often should the ground be moved? Order of ploughing? Give reasons for the seven years' course? Substances most needed for the restoration of fertility? Best manures?—714. On clover? Give an example of successful use of gypsum?—715. A fallow? When may a field lie fallow? Show how it should be managed?—716. Weathering? Its uses? Where are these salts?—717. Who have allowed their fields to be fallow? What has rendered it less necessary? What substitute? Why should the rotation vary with the soil?—718. How is a farmer to know the best course of rotation? Should it vary with the object the farmer has in view?—719. Give, upon the black-board, the 1st course laid down, with the reasons. 2d. 3d. 4th. 5th. Give a longer course, and the reasons.—720. Norfolk rotation? Give the favorite French course, with the reasons. Give an English course for clayey soils. For rich loams. What fact should essentially vary these courses when adopted in America?—721. What may sometimes be substituted for a rotation of crops?

CHAPTER XXII.—724. What is the first of the harvest? What determines the proper time for cutting grass?—725. When should it be cut for feeding to milch cows?—726. When to obtain the greatest quantity and the best quality of milk?—727. When to feed to store cattle?—728. When do grasses attain their full development? What constituents of grass are of the highest value?—729. What change takes place after flowering?—730. Why not let the seed ripen?—731. How is grass cut? What is the advantage of the mowing machine?—732. What further advantage is there? How is grass treated after being cut? What is the hay tedder?—734. How is it gathered? What is the advantage of the horse-rake?—735. On what does the time required to cure grass depend? When may it be got in?—736. What is the effect of over-drying?—737. In what does the true art of hay-making consist?—738. How may injury in the mow be prevented?—739. What is the result of experience in drying hay?—740. When should clover be cut and how cured? When may it be put in?—741. How does clover compare with other hay? For what is it most valuable?—742. How may it be saved from injury?—743. When should lucerne be cut and why?—744. When is the proper time to cut wheat and rye? When then? How are they cut?—745. How are oats and barley cut?—746. When should Indian corn be gathered? What is the custom?—747. When should potatoes be dug and how? What is the result of exposure to the sun?—748. What is the effect of exposure to the sun on the surface while growing?—749. When are turnips harvested and how?—750. When should carrots be taken out of the ground and how?—751. What is said of mangolds?

CHAPTER XXIII.—752. What is disease? How is it brought about?—753. What is a predisposing cause of disease?—754. What is an exciting cause?—755. What are some of the diseases often ascribed to parasites?—756. What is mildew?—757. What is the appearance of wheat mildew? What plants does the white mildew attack?—758. How does white mildew first appear? How does it extend?—759. What is the remedy?—760. Is wheat mildew the same as that of the vine?—761. What varieties of wheat are most affected with it?—762. What gives it the name of rust?—763. On what soils is wheat most liable to be attacked?—764. What is the remedy?—765. When may salt be applied?—766. What is the disease called smut?—767. How does it affect the grain?—768. On what soils is smut most common?—769. How may its presence in wheat be known?—770. What are the remedies?—771. How may the seed be prepared?—772. What is the disease called canker?—773. What is this disease caused by?—774. What is its peculiarity?—775. How is it propagated?—776. What is the condition of the diseased grain at the time of ripening?—777. How does it appear in threshing?—778. How may it be guarded against?—779. How may a wash be prepared for the seed?—780. What is blight?—781. What is ergot?—782. What is it caused by?—783. Where does it prevail?—784. What is the remedy?—785. How are trees often injured?—786. When may fruit trees be pruned?—787. How should wounds on trees be treated?—788. What are some of the most injurious insects?—789. What are cut worms?—790. How may they be destroyed?—791. What other insects destroy them?—792. What is said of ground beetles?—793. What of ichneumon flies?—794. How may the apple-tree caterpillar be guarded against? When are the eggs laid and how do they appear?—795. How may trees be protected from the canker worm?—796. What about the codling moth? Remedy?—797. What does the curculio attack? How may the bitten fruit be known? What prevention is named?—798. What of the apple-tree borer? How does it enter the tree?—799. When are the eggs hatched? How is it checked?—800. What are the remedies?—801. What does the striped beetle attack? How may its ravages be prevented?—802. How may squash bugs be destroyed?—803. What of the onion maggot?—804. Of the wheat midge?—805. The dor bug?—806. Of the locust-tree borer? How is it destroyed?—807. Of the rose bug?—808. How may the spring beetle be recognized? Of what is it the parent?—809. What of the striped potato beetle? Remedy?—810. The oak pruner? How prevented?—811. Of the meal worm?—812. Of scale insects on the bark? Remedy?—813. Of the chinch bug? How destroyed?—814. The army worm? Its enemies? How checked? Their history?—815. What of plant lice? On wheat?

CHAPTER XXIV.—816. Of what does the stock of the farm consist?—817. Objects of keeping horned cattle?—818. How divided?—819. How many distinct breeds in this country?—820. Their qualities?—821. Of the Ayrshires? Why kept? Their milk?—822. Of the Jerseys?—823. What of the Short-horns?—824. Where did they originate?—325. What of the Devons?—826. Of the Herefords?—827. What of the common stock?—828. Why should only good stock be kept on the farm?—829. On what does success depend?—830. Nutriment animals require ?—831. Difference in a full grown and a growing animal ?—832. Treatment of young stock ?—833. Common mistake ?—834. Consequence ?—835. Treatment of cows ?—836. What is of most importance ?—837. What of moist food in winter ?—838. How fed to produce the greatest quantity ?—839. For the best quality ?—840. How to feed for cheese making ?—841. What of the manner of milking ?—842. What is the best form of the animal to fat ?—843. How should fattening animals be treated ?—844. State an experiment on sheep.—845. How may the greatest quantity of manure be made ?—846. Difference in ground and unground food ?—847. How is a fat animal prepared for the butcher ?—848. What is the average loss or offal ?—849. Qualities of working cattle ?—850. Comparison of ox and horse labor ?—851. How are horses classified ?—852. Effect of standing in dark stables ?—853. How treated ?—854. What of barns and stables ?—855. How is good ventilation secured ?—856. Of the temperature of stables ?—857. What is said of treatment of animals ?—858. What of the breeds of sheep ?—859. Most profitable for special localities ?—860. Comparative profit of raising mutton ? Of cost of fences ?—861. Of shelter ?—862. State an experiment in keeping sheep ?—863. What is said of feeding ?—864. Amount of food consumed ?—865. How may sheep be protected ?—866. Of the breeds ? On what will the choice depend ?—867. What of the food of swine ?—868. What of poultry ?—869. What of their food in winter ?—870. What of the varieties of fowls ?

Chapter XXV.—871. On what will success depend ?—872. What of choice of location ? Advantage of good land ?—873. What is said of the location of buildings ?—874. Of the construction of fences ?—875. Of the economy of implements ?—876. Of buying too many ?—877. Of the more expensive ones ? How may they be owned ?—878. How are farm tools injured ?—879. What is said of the tool room ?—880. What mistake is often made ? On what does profit depend ?—881. What is the result of trying to do too much ?—882. What is said of protecting crops ?—883. Of the care of growing corn ?—884. What is true of every crop ?—885. A source of great loss ?—886. How is the farmer to keep up the fertility of his land ?—887. How are knowledge and skill required ?—888. How is a compost formed ?—889. The most direct way of increasing fertility ?—890. Economy of green fodder ? Of manure ?—891. Of losses from badly wintering stock ?—892. What of waste land along walls ?—893. Of allowing stones to lie in piles on the lot ?—894. What is said of a garden ? Economy of it ?—895. Of a hot-bed ?—896. When is it made ?—897. How is the frame constructed ?

—898. How are the sashes put on?—899. What exposure is best? When is the bed started?—900. Best heating material? How put in? —901. How may the heat be protracted?—902. Management?—903. How is the seed sown?—904. How may too great heat be avoided?—905. How else?—906. How may the temperature be determined?—907. What caution is required?—908. How may cucumbers, &c., be planted? —909. What plants may be started in the hot-bed?—910. What hot-beds are most easily regulated?—911. What is said of fruit culture?—912. Of young trees? Of other crops?—913. How is a too thrifty growth checked? How is the growth promoted? What of spading up around trees?—914. What of pruning? Best time?—915. When should apples and pears be gathered? Effect of ripening on the fruit? —916. What is said of the strawberry?—917. Of the raspberry and blackberry?—918. Of the gooseberry? Value of mulching?—919. When should grapes be set? When pruned? Object the first year or two?—920. Effect of ornamental trees?—921. When should wheat and other grains be harvested?—922. Effect of ripening?—923. What is said of oats?—924. Of keeping correct accounts?—925. What accounts should be kept? Result?

Chapter XXVI.—926. What of the importance of good management in the house?—927. Of the dairy?—928. Of the care of milk?—929. What is milk? Its composition?—930. The proportions of the various constituents?—931. How does it appear under the microscope? What are the globules? Their size?—932. Weight of milk? Effect of heat and cold? What change takes place when at rest?—933. What is the cause of a whitish appearance in cream?—934. Why is churning necessary? What sometimes takes place?—935. Effect of heat on milk?—936. What is the percentage of cream? Which is lighter, rich or poor milk?—937. Temperature of new milk? Depth in the pan? Temperature of the dairy-room?—938. Why is the strictest cleanliness necessary?—939. What of the first cream that rises? How may the best cream and butter be obtained?—940. Of the dairy house?—941. How should the room be located in the house? What of cellars?—942. Of the atmosphere near cellar bottoms? Where should milk stand to facilitate the rising of cream?—943. Describe a convenient form of milk stand.—944. How long should milk stand for cream? What of churning? How may cream be kept?—945. What is said of butter? At what temperature of the cream is the best quality of butter procured? How much does the temperature rise in churning? How may the cream be brought to the proper temperature?—946. What of churning?—947. What form of churn is mentioned? What is it that brings the butter?—948. What is the best way of working butter? What of the sponge and cloth? What of butter made in this way?—949. How might a simple butter-worker be made?—950. How is butter prepared for market? Effect of over-salting?—951. How may new boxes be prepared so as not to taint butter?—952. How might both nice butter and cheese be made?—953. What is cheese made from?—954. From what substances may it be made? How does the acid act?—955. What is the process of making cheese?—956. What is said of the rennet? What of the pressing?—957. What of the quality of dairy produce?—958. What is said of bread making?—959. Of the flour?—960. Composition of wheat?—961. What is starch?—962. What of gluten? How may it be washed from dough?—963. What is bran?—964. Effect of water on flour? What is dough?—965. Effect of yeast on dough?—966. What makes bread light and spongy? What takes place when dough is put into the oven?—967. What is yeast?—968. What changes does it produce? What checks rising in the oven?—969. What change takes place in the bread? Is stale bread actually drier than new? Which is the more healthful?—970. How does the gluten retain water?—971. How does the value of flour depend on the amount of gluten?—972. State an experiment with different flour?—973. How much water does flour in its natural state contain? How much bread will a hundred pounds of flour make?—974. How does the bran compare with flour?—975. What of rye flour? Rye bread?—976. Why is wheat flour preferred to rye?—977. Of Indian meal?—978. The most common modes of cooking meats? What does beef lose in boiling? Mutton?—979. What per cent. of water in beef? What in beef corresponds with the gluten of bread? In what does the difference consist?—980. What is fibrin? How is fat deposited?—981. To what is the loss in cooking due? What is the juice of meat?—982. How is it preserved?—983. How would much of the richness of meats be lost?—984. How does the case differ in preparing broths and beef tea?—985. On what does the process of soap making depend?—986. The difference between soft and hard soaps?—987. How is castile soap made?—988. What of rosin or yellow soaps?—989. On what do the cleansing properties of soap depend?—990. Why does water without soap often fail to cleanse?—991. How is home to be made attractive?—992. What is calculated to secure this end?

INDEX.

www.ingramcontent.com/pod-product-compliance
Lightning Source LLC
LaVergne TN
LVHW020233110826

845151LV00003B/915

9781425530044